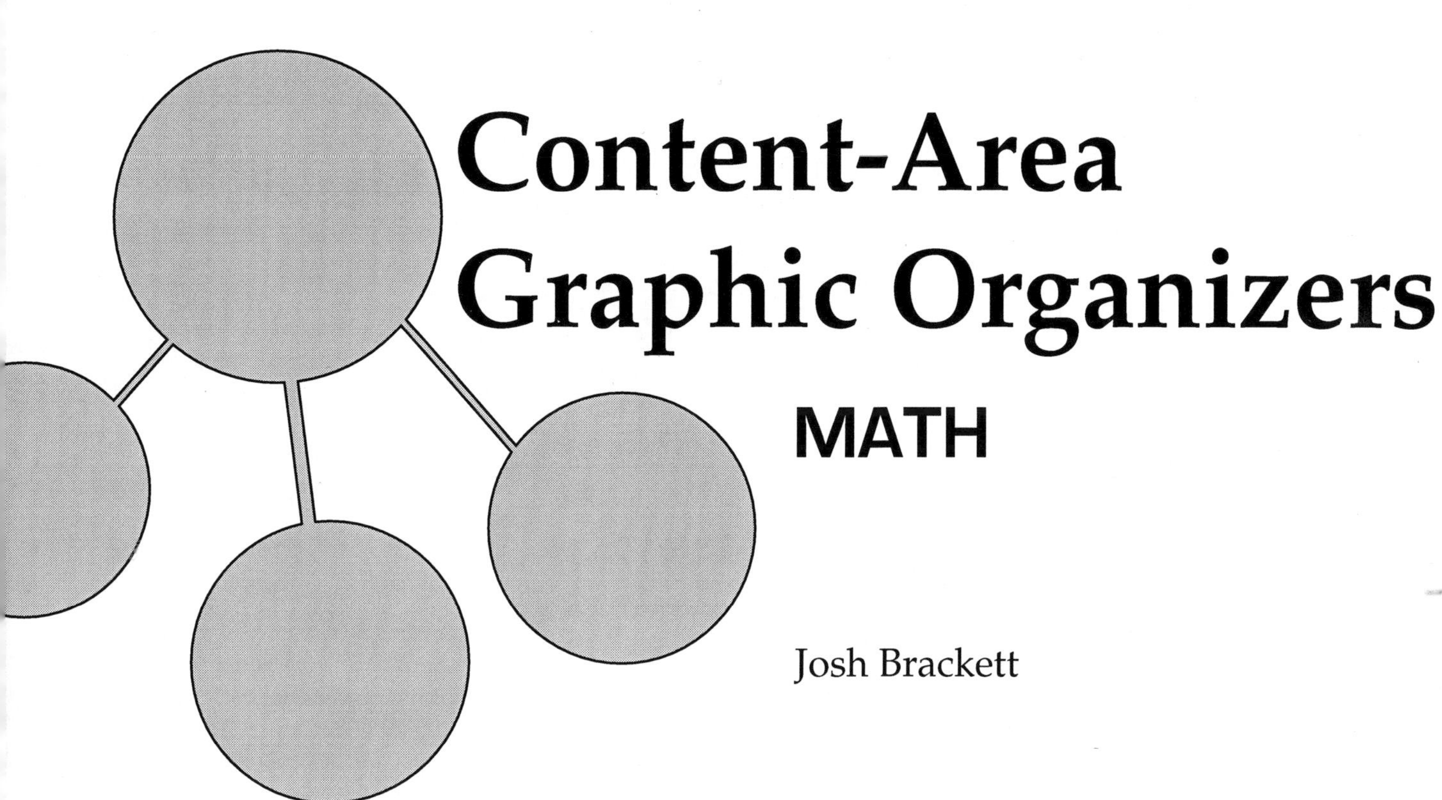

Content-Area Graphic Organizers

MATH

Josh Brackett

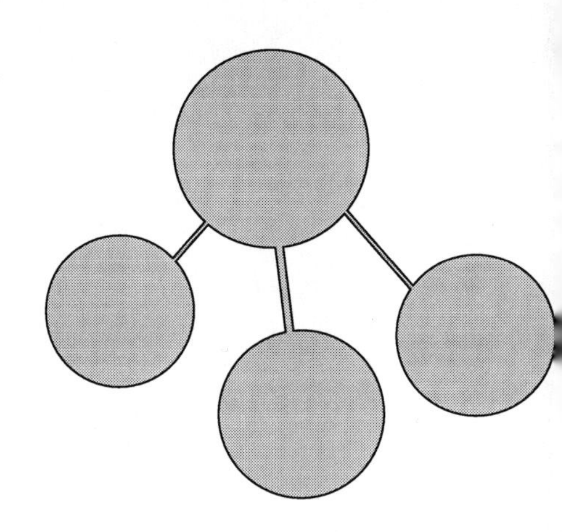

Table of Contents

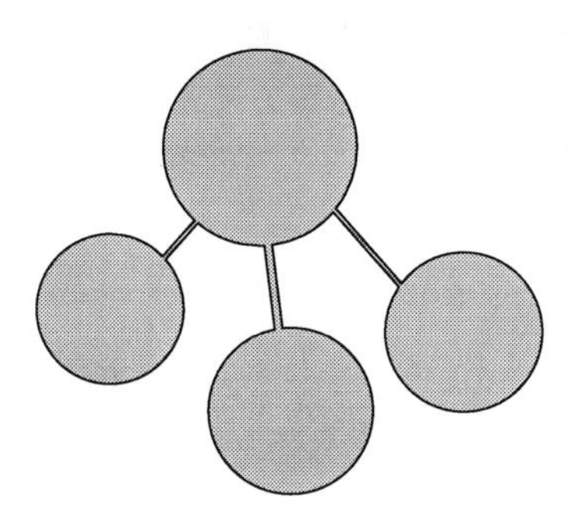

To the Teacher

Graphic organizers can be a versatile tool in your classroom. Organizers offer an easy, straightforward way to visually present a wide range of material. Research suggests that graphic organizers support learning in the classroom for all levels of learners. Gifted students, students on grade level, and students with learning difficulties all benefit from their use. Graphic organizers reduce the cognitive demand on students by helping them access information quickly and clearly. Using graphic organizers, learners can understand content more clearly and can take clear, concise notes. Ultimately, learners find it easier to retain and apply what they've learned.

Graphic organizers help foster higher-level thinking skills. They help students identify main ideas and details in their reading. They make it easier for students to see patterns such as cause and effect, comparing and contrasting, and chronological order. Organizers also help learners master critical-thinking skills by asking them to recall, evaluate, synthesize, analyze, and apply what they've learned. Research suggests that graphic organizers contribute to better test scores because they help students understand relationships between key ideas and enable them to be more focused as they study.

This book shows students how they can use some common graphic organizers as they read and write in math classes. As they become familiar with graphic organizers, they will be able to adapt them to suit their needs.

In the math classroom, graphic organizers help students:

- preview new material
- make connections between new material and prior learning
- recognize patterns and main ideas in reading
- understand the relationships between key ideas
- organize information and take notes
- review material

This book offers graphic organizers suitable for math tasks, grouped according to big-picture skills, such as storing and retrieving information; problem-solving; and communicating with others. Each organizer is introduced with an explanation of its primary uses and structure. Next comes a step-by-step description of how to create the organizer, with a worked-out example that uses text relevant to the content area. Finally, an application section asks students to use the techniques they have just learned to complete a blank organizer with information from a sample text. Throughout, learners are encouraged to customize the organizers to suit their needs. To emphasize the variety of graphic organizers available, an additional organizer suitable for each big-picture skill is introduced briefly at the end of each lesson.

Content-Area Graphic Organizers for Math is easy to use. Simply photocopy and distribute the section on each graphic organizer. Blank copies of the graphic organizers are included at the back of this book so that you can copy them as often as needed. The blank organizers are also available for download at our website, www.walch.com.

As learners become familiar with using graphic organizers, they will develop their own approaches and create their own organizers. Encourage them to adapt them, change them, and create their own for more complex strategies and connections.

Remember, there is no one right way to use graphic organizers; the best way is the way that works for each student.

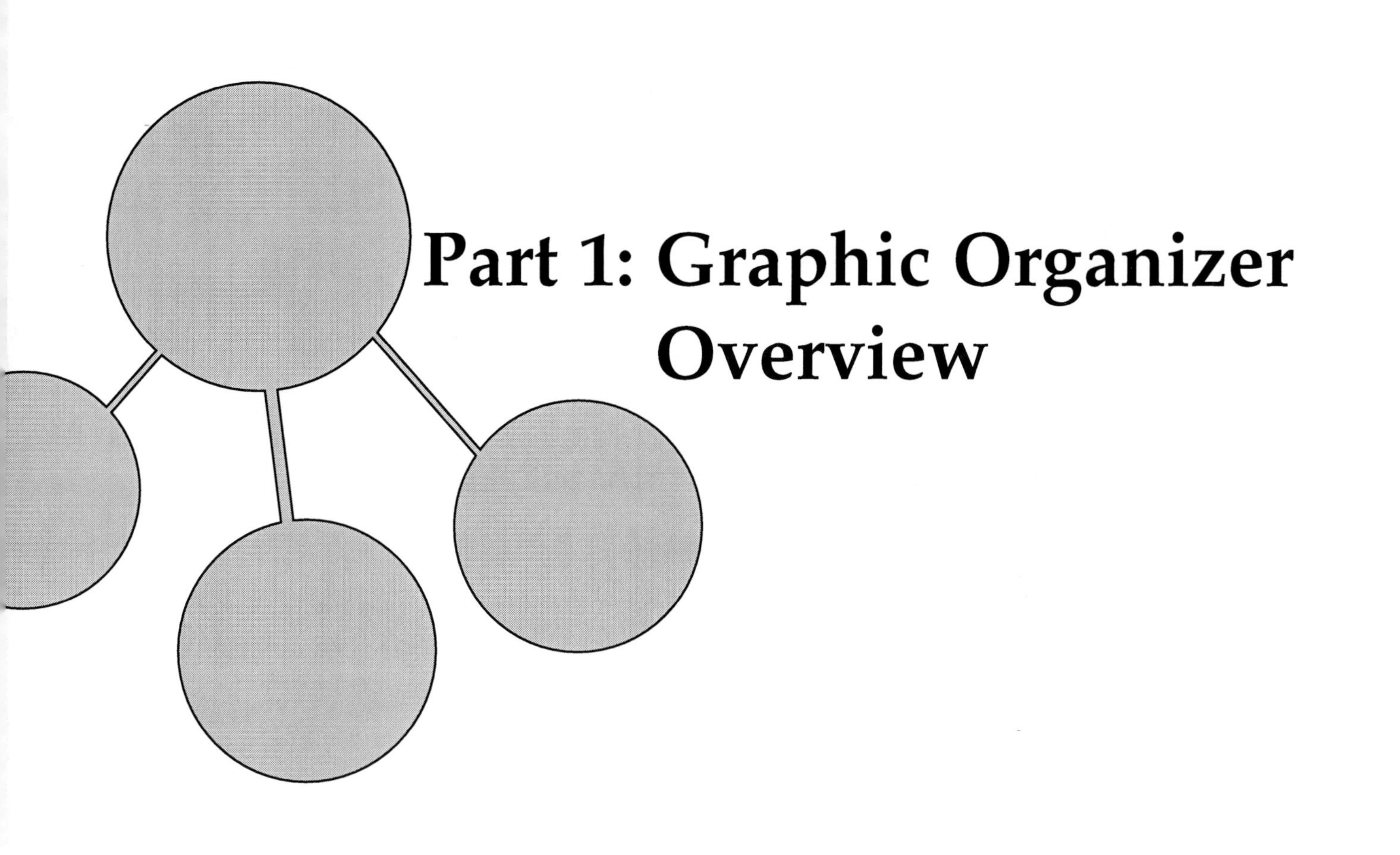

Part 1: Graphic Organizer Overview

Introduction to Graphic Organizers

You've probably heard the old saying, "A picture is worth a thousand words." Like most old sayings, it isn't always true. But in many things we do, words alone are not the best way to communicate. That's why we use pictures and, in particular, graphic organizers.

A graphic organizer is simply a special drawing that contains words or numbers. If you've ever made a web or filled in a chart, then you already know how to use a graphic organizer. In this book, you'll find that you can use graphic organizers in ways you may not have expected. And you'll find that they can make your learning a lot easier!

The power of a graphic organizer is that instead of just telling you about relationships among things, it can show them to you. A graphic organizer can help you understand information much more easily than the same information written out as a paragraph of text. For example, look at this listing of names, addresses, and telephone numbers. Use it to find the telephone number for Amanda Jones.

Alden E. Jones, 18 Milford St., Boston, MA 02118, (617) 555-8040. Alun Huw Jones, 91 Westland Ave., Boston, MA 02115, (617) 555-9654. Alvin Jones, 715 Tremont St., Boston, MA 02118, (617) 555-2856. Alvin D. Jones, 77 Salem St., Boston, MA 02113, (617) 555-2890. Amanda Jones, 111 W. 8th St., Boston, MA 02127, (617) 555-0738. Amos K. Jones, 11 Helen St., Boston, MA 02124, (617) 555-3560. Andre N. Jones, 523 Mass. Ave., Boston, MA 02118, (617) 555-0829. Andrew Jones, 168 Northampton St., Boston, MA 02118, (617) 555-0069.

In order to find Amanda's number you had to read, or at least scan, the whole text. Here is the same information presented in a graphic organizer—a table.

Name	Address	City, State, Zip	Phone
Alden E. Jones	18 Milford St.	Boston, MA 02118	(617) 555-8040
Alun Huw Jones	91 Westland Ave.	Boston, MA 02115	(617) 555-9654
Alvin Jones	715 Tremont St.	Boston, MA 02118	(617) 555-2856
Alvin D. Jones	77 Salem St.	Boston, MA 02113	(617) 555-2890
Amanda Jones	111 W 8th St.	Boston, MA 02127	(617) 555-0738
Amos K. Jones	11 Helen St.	Boston, MA 02124	(617) 555-3560
Andre N. Jones	523 Mass Ave.	Boston, MA 02118	(617) 555-0829
Andrew Jones	168 Northampton St.	Boston, MA 02118	(617) 555-0069

Which arrangement was easier to use? Most people find it easier to see the information in the table. This is because the table gives all the names in one column, all the telephone numbers in another column, and all the information about each person in one row. As soon as you know how the table is set up—the labels at the top of each column tell you—you can quickly find what you're looking for.

Graphic organizers use lines, circles, grids, charts, tree diagrams, symbols, and other visual elements to show relationships—classifications, comparisons, contrasts, time sequence, parts of a whole, and so on—much more directly than text alone.

You can use graphic organizers in many ways. You can use them before you begin a lesson to lay the foundation for new ideas. They can help you recall what you already know about a subject and see how new material is connected to what you already know.

You can use them when you are reading to take notes or to keep track of what you read. It doesn't matter what you are reading—a textbook, a biography, or an informational article. Organizers can help you understand and analyze what you read. You can use them to recognize patterns in the reading. They can help you identify the main idea and its supporting details. They can help you compare and contrast all kinds of things, from people to ideas, animals, and events.

Graphic Organizers can help you after you read. You can use them to organize your notes and figure out the most important points in what you read. They are a great tool as you review to make sure you understood everything or to prepare for a test.

You can use graphic organizers when you write, too. They are particularly useful for prewriting and planning. Organizers can help you brainstorm new ideas. They can help you sort out the key points you want to make. Graphic organizers can help you write clearly and precisely.

Think of graphic organizers as a new language. Using this new language may be a bit awkward at first, but once you gain some fluency, you will enjoy communicating in a new way.

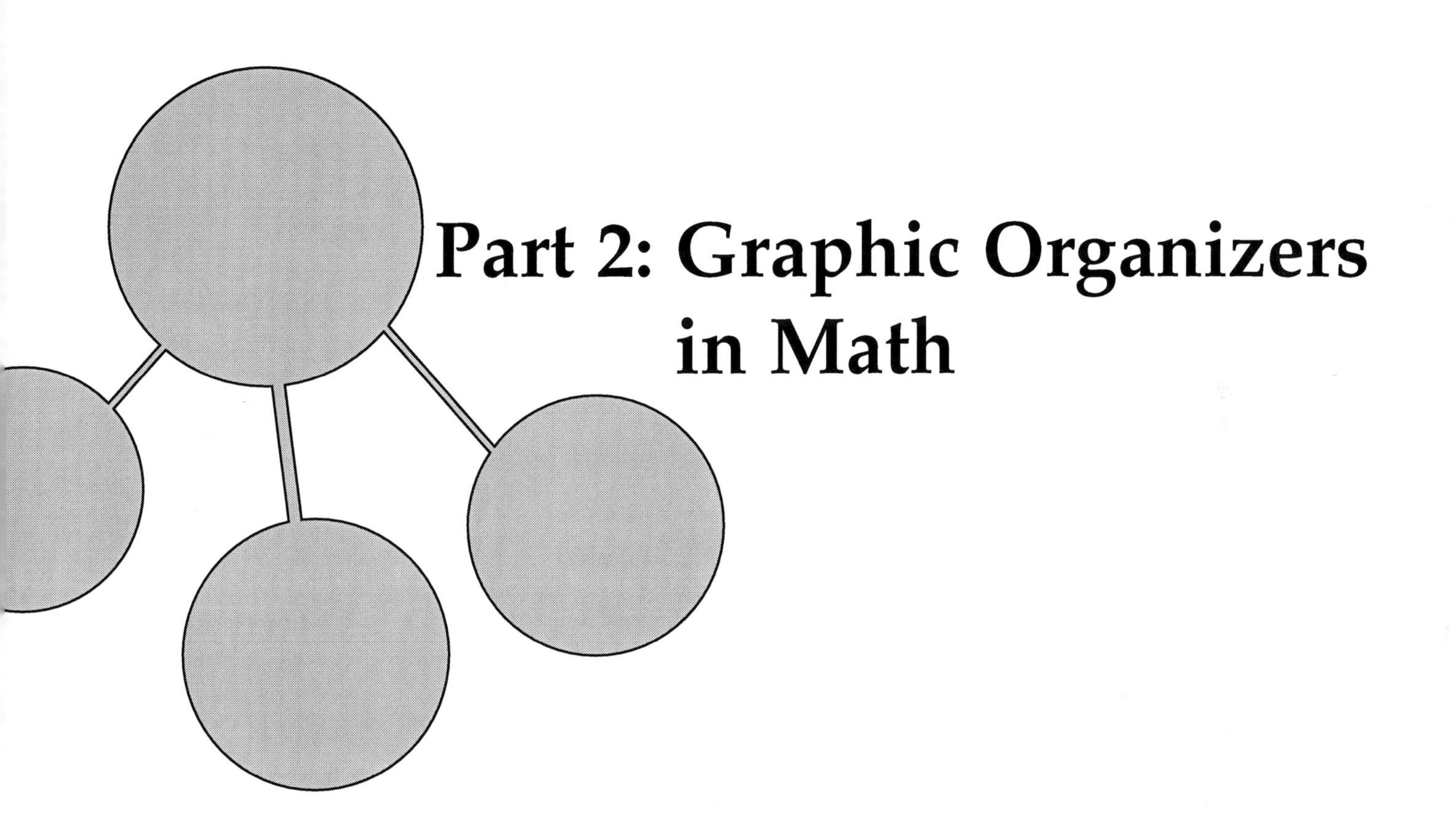

Part 2: Graphic Organizers in Math

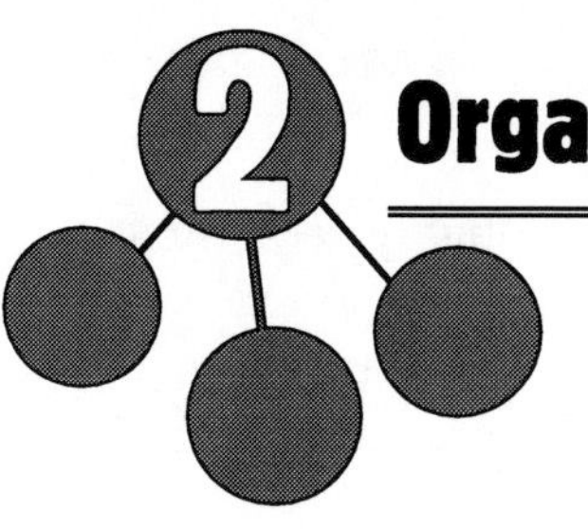

Organizing, Categorizing, and Classifying

Have you ever read an explanation and found it hard to remember the important points? Perhaps something distracted you from what the author was saying. Perhaps the material was poorly organized, or contained a lot of new information.

One of the best ways to keep track of important information is to use a graphic organizer to take notes. Graphic organizers can help you sort information into categories. This makes it easier to remember what you have read. And once you have organized the information, you can use the organizer as a reference while you are solving problems in class or for homework, or when you are reviewing for a test.

The way you organize your notes depends on the material you are reading. This lesson will present two organizers that are useful for organizing mathematical material: tables and flowcharts.

Tables are a way to show comparisons and contrasts. Flowcharts are a visual approach to laying out the steps in a process. You can use these graphic organizers to record mathematical information that you need to recall later. And once you learn how to make them, you'll find other uses for them, too.

Tables

Of all the graphic organizers in this book, the one you have probably seen most often and will use the most, both in math and elsewhere, is the table. A table is simply a grid with rows and columns. Tables are useful because information stored in a table is easy to find—much easier than the same information embedded in text.

Tables are so useful that software packages, such as Word and Excel, offer table-making capabilities. But you don't need a computer to make a table. All you need is a pencil and paper. If you make tables by hand, you can include them in your handwritten notes.

Tables in Action

Tables come in all sizes and shapes. The size and shape of a table depends, of course, on what's in it. When you're reading material that you will need to recall later, the first step is to think about how to take notes on it. Should you make a table or should you take notes in some other form? If what you're reading is a description, a narrative, or a logical argument, it may not lend itself to storage in a table. A table is essentially a list of things that have something in common with one another, and the attributes that they have in common. If, as you read, you find yourself thinking "Oh, this is a list of . . .," a table is probably a good organizer to use.

Usually, a table has a row (the horizontal part) for each item being listed. The columns (the vertical part) provide places for aspects of the listed items—the things they have in common. The places where the rows and columns meet are called cells. In each cell, we write information that fits both the topic of the row—the thing being listed—and the topic of the column—the aspect being examined. To create a table, we make rows and columns to fit the number of items and attributes.

For example, look at this reading about Platonic solids. The first sentence tells us that there are five of them and names some attributes of the first one. Is this going to be essentially a list of items that all have certain attributes? A quick scan of the text tells us that it is. So we'll take notes on it in the form of a table.

Platonic Solids

There are five Platonic solids. The first is called the tetrahedron, which is Greek for "four faces." Each face is an equilateral triangle. It has 4 vertices and 6 edges, with 3 triangular faces meeting at each vertex.

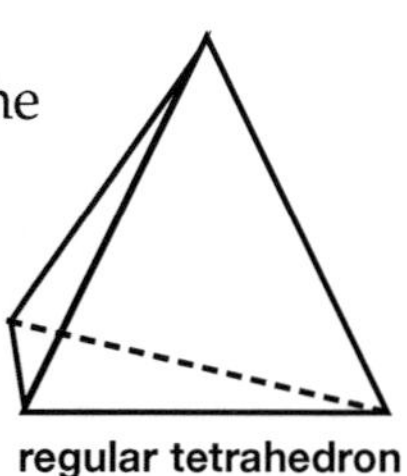

regular tetrahedron

The second Platonic solid is the cube. It has 6 faces, each of which is a square. It has 8 vertices and 12 edges, with 3 faces meeting at each vertex.

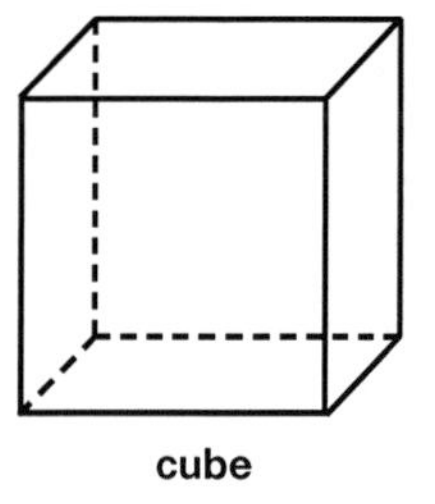

cube

The third Platonic solid is the octahedron, which in Greek means "eight faces." Like the tetrahedron, the octahedron's faces are equilateral triangles. It has 6 vertices, with 4 triangles meeting at each vertex, and, like the cube, it has 12 edges.

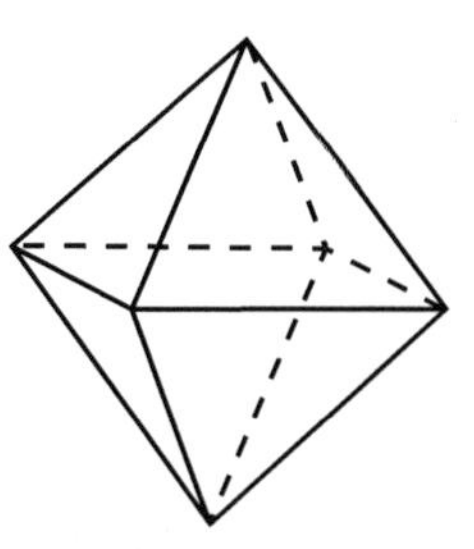

regular octahedron

Next comes the dodecahedron, a 12-faced solid. The faces of the dodecahedron are regular pentagons, 3 of which meet at every vertex. There are 20 vertices and 30 edges.

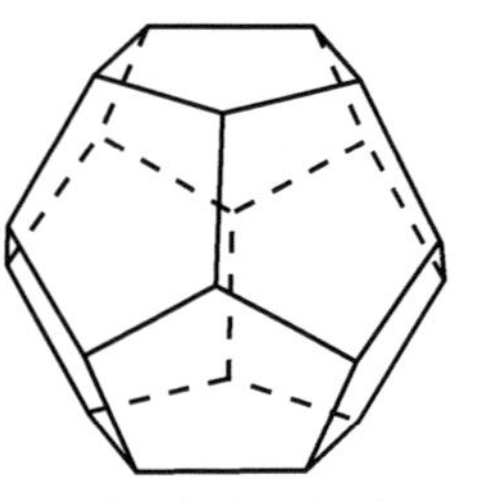

regular dodecahedron

The last Platonic solid is the icosahedron, a 20-faced solid. Each of the 20 faces of the icosahedron is an equilateral triangle. It has 5 triangles that meet at each of 12 vertices, and there are 30 edges.

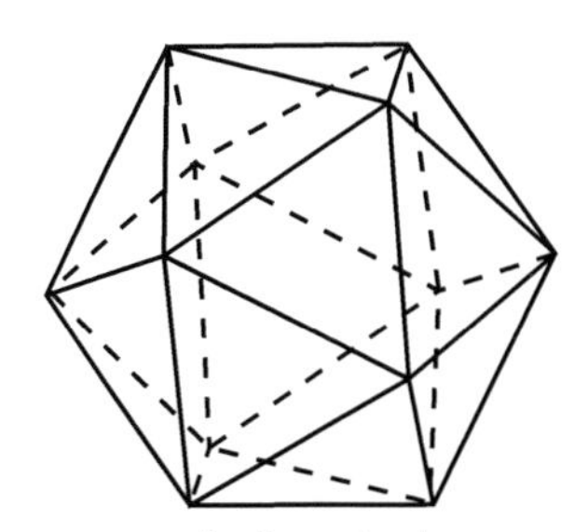

regular icosahedron

Before we can make a table, we have to know what the rows and columns are going to be. Since the text is essentially a list of the five Platonic solids with information about each one, the table will have a row for each solid, or five rows.

What are the columns going to be? To answer this we go back to the text. What kinds of information does it give us about each Platonic solid? It gives us the number of faces, what regular polygon each face is, the number of vertices, the number of faces that meet at each vertex, and the number of edges. That tells us both how many columns we need and what the column heads will be. We need a total of seven columns, one for the name of the solid and one for each piece of information.

Now we can draw the table, adding the appropriate number of rows and columns. Write the column heads at the top of each column. Then write the name of each Platonic solid at the start of each row.

Finally, we can go through the text again to find the information for each cell.

Here's the finished table.

Platonic Solid	**Faces**	**Polygon**	**Vertices**	**Faces Meeting at Each Vertex**	**Edges**
tetrahedron	4	equilateral triangle	4	3	6
cube	6	square	8	3	12
octahedron	8	equilateral triangle	6	4	12
dodecahedron	12	pentagon	20	3	30
icosahedron	20	equilateral triangle	12	5	30

The information is much more compact in table form than in text. It's also more accessible. If you need to look up a characteristic of a particular solid, you can find it easily. Also, a table makes it easier to compare the properties of the different Platonic solids. For example, to see how the number of edges increases as the number of faces increases, you can look from the faces column to the edges column.

You can make tables on a computer using software programs, such as Word or Excel. However, if you are taking notes on your own reading, it is often quicker and easier to make a table by hand. That way you can include tables on pages of your handwritten notes.

No matter how you prepare the table, follow the same basic steps.

1. Decide what the rows and columns are going to be—the items being listed and the attributes of each one.

2. Draw a table with the appropriate number of rows and columns.

3. Write the items being listed at the start of each row. Write the aspects being examined at the top of each column.

4. Fill in the cells with information that fits both the row and the column.

Application Read the text below about area formulas. Then use the table on page 13 to take notes on the text.

Area Formulas

The area of a polygon is the number of square units of measure needed to cover it completely. A rectangle is the simplest polygon with which to measure the area. The number of square units a rectangle contains is the product of the length of one side by the length of a side perpendicular to it. We call one of these sides the base and the other the height. So the formula for the area A of a rectangle is its base b times its height h. $A = bh$.

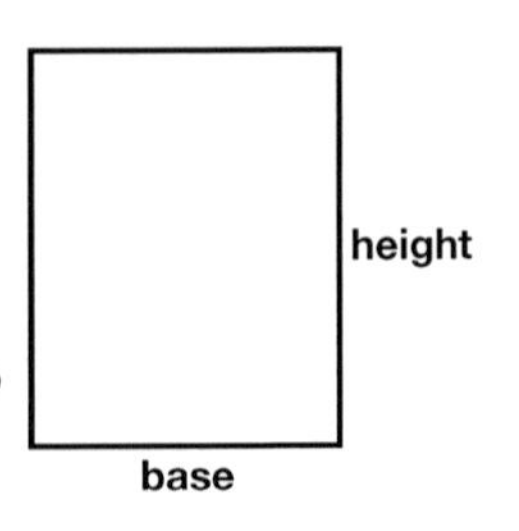

A square is a special kind of rectangle. The formula for the area of a square is still $A = bh$. But in a square, the base and height are equal. So we often call the base and height sides s and write the formula as $A = s^2$.

The formula for the area of a parallelogram is also the same as for a rectangle. The only difference is that the height of a parallelogram is not the length of a side. It is the length of a line segment perpendicular to the base from the base to the opposite side (like the side of a rectangle). As long as you define height in that way, the formula for the area A of a parallelogram is its base b times its height h. $A = bh$.

Once you know the formula for the area of a parallelogram, the formula for the area of a triangle follows logically. By drawing a diagonal line, you can divide any parallelogram into two congruent triangles—triangles that have the same size and shape.

Since the two triangles are congruent, the area of each triangle is half of the area of the parallelogram. Therefore, the area of a triangle is half its base b times its height h. $A = \frac{1}{2}bh$.

We can derive the formula for the area of a trapezoid in a similar way. When we draw a diagonal, we divide the trapezoid into two triangles that are not congruent. Notice that they have different bases—we'll call them base 1 (b_1) and base 2 (b_2)—but the same height. The area of the trapezoid is the sum of the areas of the two triangles formed by the diagonal. $A = \frac{1}{2}b_1h + \frac{1}{2}b_2h = \frac{1}{2}(b_1 + b_2)h$.

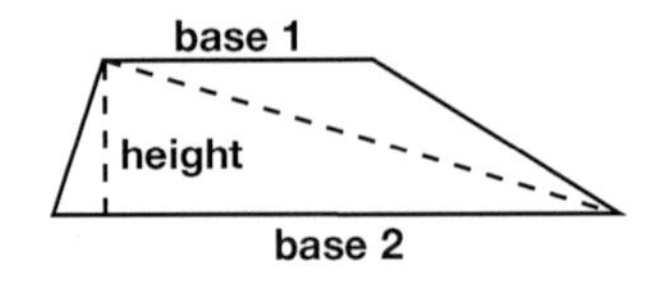

Table Use the blank table below to summarize the information from the text on page 12. Add or delete columns and rows as needed. Remember the steps for preparing a table: First, decide on the information to go in the rows. Next, see what information needs to go in the columns. Then write in the column heads and the names of the things being listed. Finally, fill in the information from the text.

Flowcharts

Flowcharts are graphic organizers that show the steps in a process. Flowcharts can be very simple—just a series of boxes with one step in each box. However, there is also a more formal type of flowchart. These flowcharts use special symbols to show different things, like starting and stopping points or points where decisions must be made. These symbols make flowcharts especially useful for showing complicated processes. Whenever a process needs to show several different options—"if this happens, then you follow this step, but if that happens, then you follow another step"—a flowchart is probably the best way to chart the steps.

Flowcharts are often used in science and business. They can show all kinds of different processes: how factories make products, how computers process data, and so on. In math, you can also use flowcharts to show many different processes, from adding a column of numbers to using formulas. You can use them to take notes as you read or as your teacher explains a new concept. You can use them as a guide to a process when you are solving problems, either in class or at home. And you can use them as a reminder when you study for a quiz or a test.

Each step in a flowchart is written in a box. The boxes are connected by arrows to show the sequence of steps. The boxes aren't all rectangular; different shapes are used to indicate different actions. The shapes and symbols are a kind of visual shorthand. Whenever a certain symbol is used, it always has the same meaning.

Flowchart Symbols

Circles and ovals show starting and stopping points. They often contain the words "start" or "stop." The "start" box has no arrows in and one arrow out. The "stop" box has one arrow in and no arrows out.

Arrows show the direction in which the process is moving.

Diamonds show points where a decision must be made or a question must be answered. The question can usually be answered either "yes" or "no."

Rectangles and squares show steps where a process or an operation takes place.

Parallelograms show input or output, such as writing or printing a result or solution.

Flowcharts in Action

To make a flowchart, you have to think through the process carefully. How does the process start? How does it end? What happens in between? In what order? Do any steps call for making decisions or choices? Do any steps have more than one possible result? Try to break down the process into the smallest possible parts. Once you have done this, you can use the symbols given on page 14 to create the flowchart.

Let's look at a common mathematical process: addition. We'll make a flowchart for the process of adding two-digit numbers.

The first step in making a flowchart is identifying the "start" point of the process. When adding two-digit numbers, where do we start? We write the numbers in a column so that the digits in the tens place and in the ones place are lined up. How can we categorize this step? Is it a decision, an output, or an operation? It's something we do to the numbers, so it's an operation. We can draw in the "start" oval, then a rectangle for the first step. The text in the rectangle should be as short as possible. We don't need to write full sentences—just enough to get the information across.

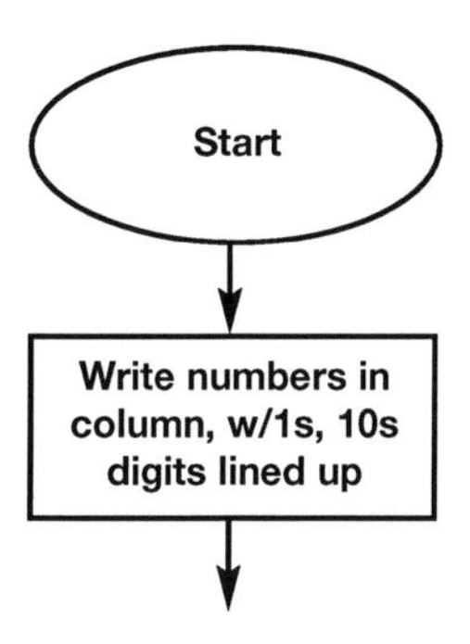

Now, identify the next steps in the process. Look for places where more than one thing can happen. These steps will go in diamonds, to show they call for questions or decisions. Try to phrase questions so the answer can be "yes" or "no." Remember to add a box for both results. Keep going until you finish the process.

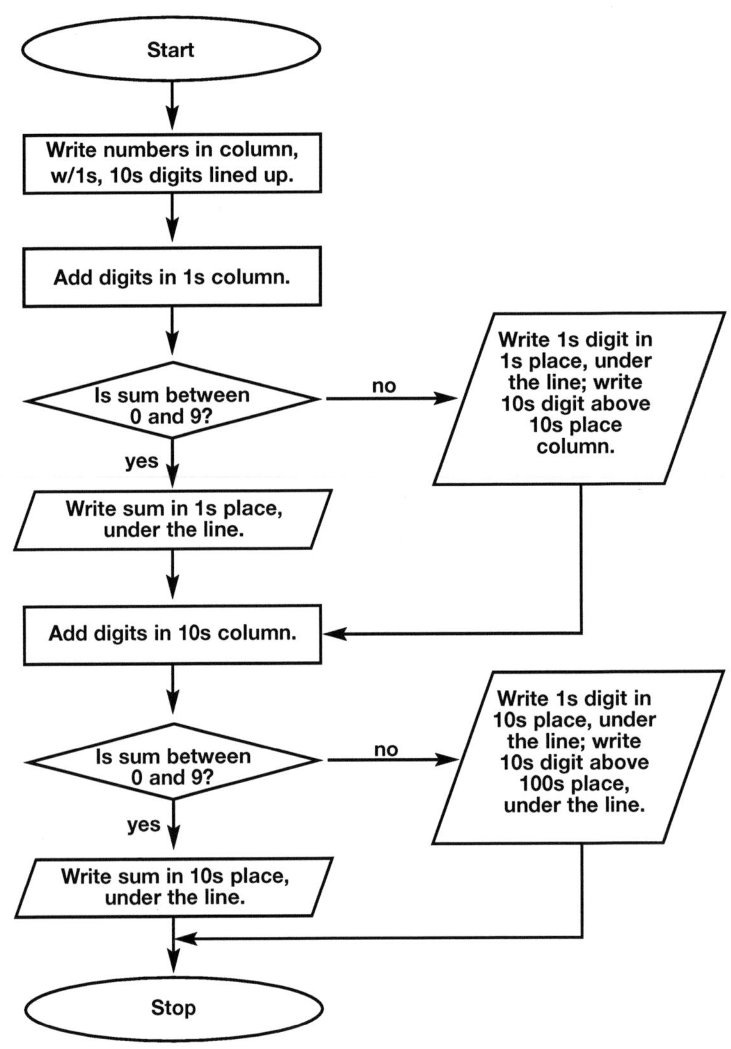

Let's review the steps for creating a flowchart.

1. Think through the process, and break it down into steps.

2. Decide whether each step is an input, an output, an operation, or a point where a decision must be made.

3. Write decision points as questions that can be answered "yes" or "no." Include options for both answers.

4. Use flowchart symbols to chart the process, beginning with a shape labeled "start" and ending with a shape labeled "stop."

Application Here are the rules for the order in which operations should be done when evaluating arithmetic expressions. Use the lines below to rewrite the rules as a process. Remember to look for points where more than one thing can happen. Include both possible results in your process. When your written version of the process is complete, use it to fill in the flowchart on page 18.

Rule 1: First, perform any calculations inside parentheses.

Rule 2: Next, perform all multiplication and division, working from left to right.

Rule 3: Last, perform all addition and subtraction, working from left to right.

Process for Evaluating Arithmetic Expressions

Flowchart Use the blank flowchart below to chart the process on page 17. Add, delete, or change boxes and lines as needed.

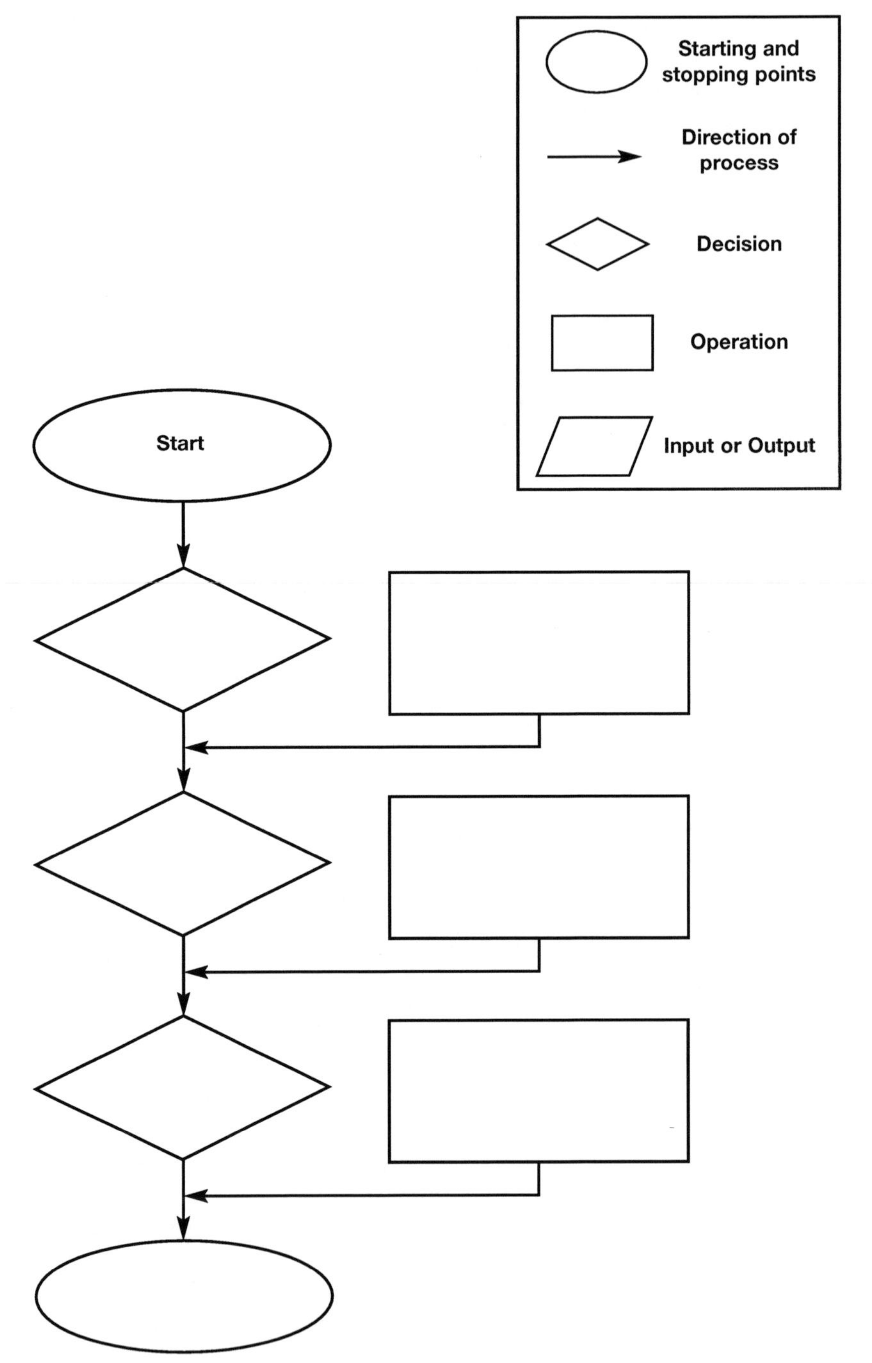

Web We have looked at two types of organizers in this lesson, but there are many other ways to organize information. Here is another organizer you could use for taking notes. Write the main idea in the center circle. Write details in the other circles. Draw lines to connect related topics. Add or delete lines and circles as needed.

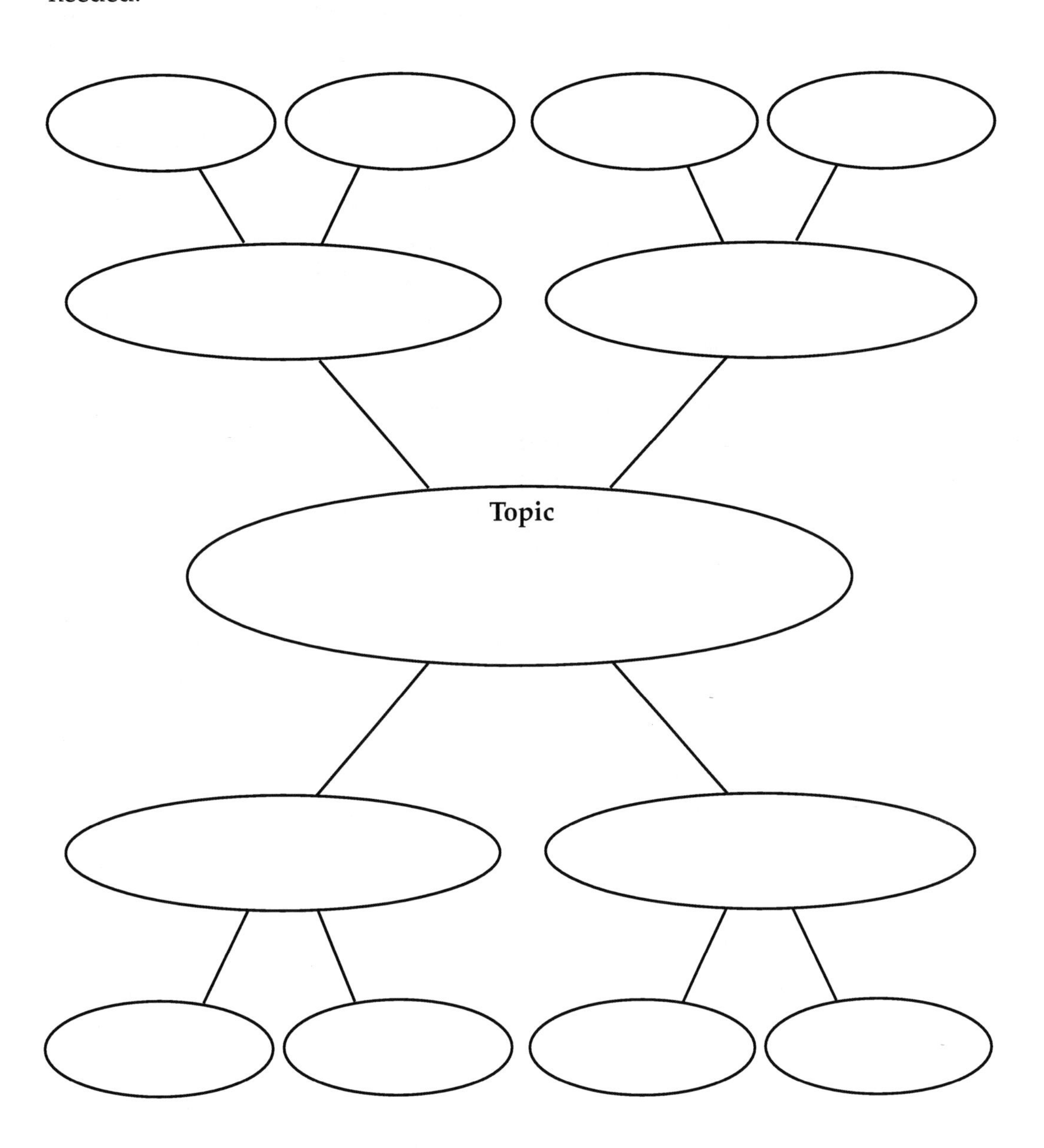

Problem Solving

Have you ever read a math problem over and over and not known where to start? Sometimes it can help to look at the problem in a different way. The graphic organizers in this lesson will help you do just that. They will show you how to turn words and numbers into drawings and diagrams.

Math problems presented in text alone—word problems—are often harder to solve than problems given in numbers or with geometric figures. There are several reasons for this.

First, a word problem may be presented in the language of the application—the speed of a moving car, the height and width of a wall—instead of in the language of math. To use your mathematical knowledge to solve the problem, you have to translate it into mathematical notation.

Second, text is one-dimensional. The words come at you one at a time in a certain order. That's why text is a good medium for telling stories but may not be the clearest way to present a math problem. To find the information that is the key to solving a problem, you may have to read through the text several times.

And finally, word problems may include information that isn't needed to solve the problem. You need to evaluate all the information to see if you need it or not. When you can't go directly from the text of a problem to a solution, a graphic organizer can be a useful intermediate step. A graphic organizer can help you see the problem in two-dimensional form. Translating a problem from text to a picture lets you see it in a new way—a way that often helps you solve the problem.

In this lesson, we show each organizer being used with one type of problem. These are problems that organizers are particularly useful for solving. However, that doesn't mean that these are the only problems with which you can use these organizers. Some problems may even be easier to solve with a combination of organizers. Remember, there is no right way or wrong way to use graphic organizers—just the way that works best for you.

Number Lines

You have probably seen and used number lines many times. In its simplest form, a number line is any line that uses equally spaced marks to show numbers. A ruler is a number line. So is a thermometer, or a measuring cup. They all show numbers and have marks evenly spaced along them.

A number line is like a map. Just as the map of a country has a dot for every major city, a line has a point for every real number. A number's place on the line shows how it relates to other numbers on the line. As numbers move to the right, they have larger values. As they move to the left, they have lower values. In the diagram below, 1 is to the right of –3. This means that 1 has a greater value than –3.

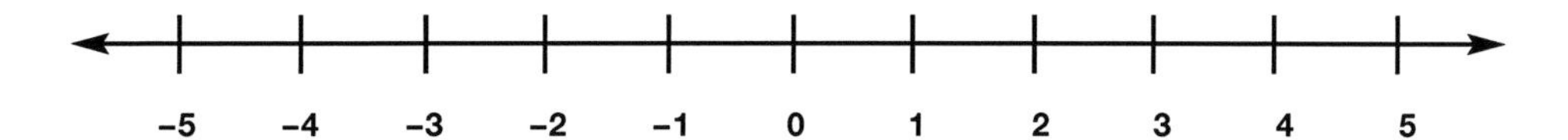

The number line above included zero. Lines like this are useful when we are working with positive and negative numbers. But number lines don't have to include zero. We can start a number line anywhere along the line. We can even create a number line when we don't know the values of the points on the line. In algebraic notation, we use letters to represent numbers whose values we don't know. In the same way, we can use letters to label points on a line that represent unknown numbers.

Even if we don't know the values of the numbers, the rules of number lines still apply. Numbers to the right have greater values than numbers to the left. In the diagram below, c is to the right of b, so we know that c has a greater value than b.

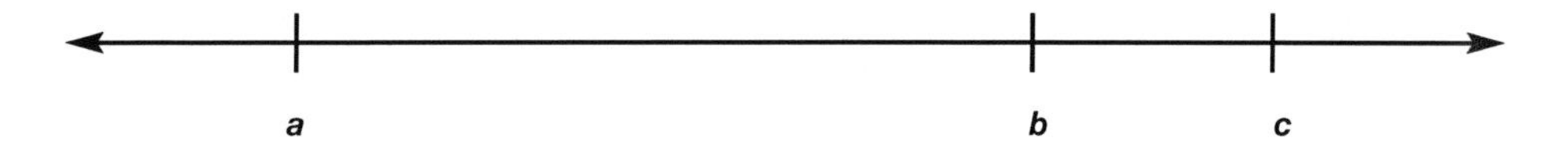

You can use number lines to visualize equalities and inequalities, positive and negative numbers, and measurements of all kinds.

Number Lines in Action

You can use number lines to "map" math problems, especially ones that involve negative numbers or distances. Let's use a number line to solve a word problem that involves distances.

> Mada is driving from Sarah's house to Chloe's. When she has gone 20 miles beyond halfway, she will be 5 miles from Chloe's house. When she arrives at Chloe's, how far will Mada have driven?

How can we figure out the answer? With word problems, the first step is always making sure we know what the problem is looking for. Restating the question sometimes helps. The real question here is, "What is the distance between Sarah's house and Chloe's house?" We are looking for the distance between two points.

Since we are looking for a distance, a number line could be a good way to map the problem. What are we going to put on the line? We don't know how far apart the two houses are. But we can use variables to show them as points on the line. Let's use S for " Sarah's house" and C for "Chloe's house." We can mark those two points on the line. Since we don't know how far apart they are, we won't mark any points between them.

The problem says, "When she has gone 20 miles beyond halfway . . ." Although we don't know the distance from Sarah's to Chloe's, we know where the halfway point is on the number line: it's halfway between point S and point C. Let's mark that point. We will label it $\frac{d}{2}$ because it represents half the distance *d* between Sarah's and Chloe's.

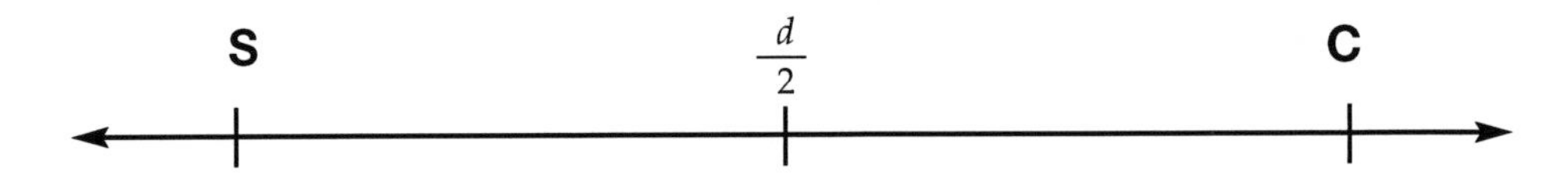

Now let's mark the point 20 miles beyond halfway. We don't know exactly where it is, but we know it's somewhere between $\frac{d}{2}$ and C. We'll call it $(\frac{d}{2} + 20)$. Since we know this point is 20 miles from $\frac{d}{2}$, moving toward C, we can put that distance on the line.

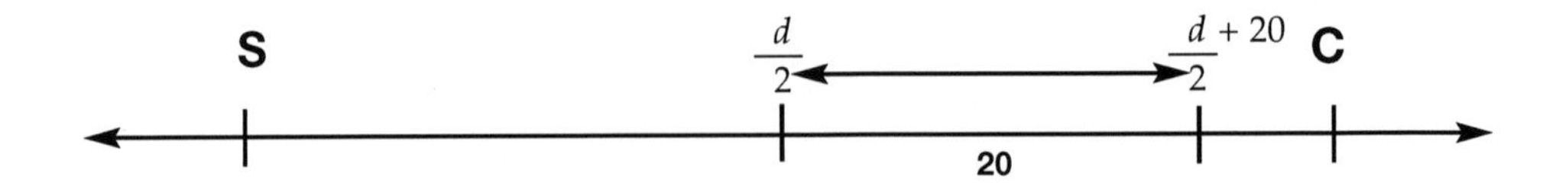

The problem says that $\frac{d}{2}$ + 20 is "5 miles from Chloe's house," so we can put in that distance.

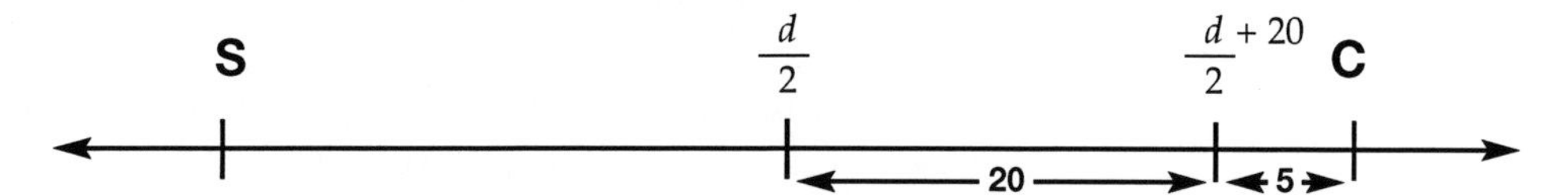

Looking at the number line, we see that the distance from $\frac{d}{2}$ to C is 20 + 5, or 25 miles. This means that the distance from the halfway point to Chloe's house is 25 miles. To find the full distance from Sarah's house to Chloe's, we just have to multiply this by 2: $25 \times 2 = 50$. The distance from Sarah's house to Chloe's house is 50 miles.

Let's review the steps in using a number line to solve word problems.

1. Analyze the problem to see if it is suitable for a number line. Problems that involve negative numbers and distance can often be solved using number lines.

2. Decide what the problem is really asking for. If necessary, restate the problem.

3. Draw a number line, and mark any information you know on the line.

4. Use the number line to find the missing information.

Application For each problem below, decide what the problem is asking you to find. Write this in question form on the lines above each number line on page 25. Then use the number lines to solve the problems.

1. Rafi is fascinated by weather. He plans to become a meteorologist. Rafi has joined an online weather network. Network members track their local weather statistics. They post their findings to a web site, where each member's findings are pooled with those of other members. Over time, this gives network members a detailed picture of weather patterns in different parts of the world.

   The statistics being measured include maximum temperature, minimum temperature, and daily temperature range. Network members post each of these measurements each day.

   On January 8, Rafi recorded the temperature at five different times over the course of the day. At 7:00 A.M. it was –3° F. At 11:00 A.M. it was 4° F. At 3:00 P.M. it was 2° F. At 7:00 P.M. it was –6° F. At 11:00 P.M. it was –12° F. What figures should Rafi post to the web site?

2. Jordan is a customer service representative for Icarus Airlines. The weekends before and after school vacations are busy times. The terminal is always filled with people who want to fly standby—to get seats on flights for which they don't have reservations. And planes are often overbooked, so unless passengers volunteer to take a later flight, people get bumped from their flights.

   On the Sunday after spring break, Jordan has been juggling standby passengers, bumped passengers, and volunteers for other flights. Another flight is preparing to board right now. The plane seats 256 passengers. A total of 265 seats on the plane have been booked. The Lees, a family of 5, are waiting by Jordan's desk, hoping to be able to fly standby on this flight. At this stage, 254 of the passengers who had booked tickets have checked in. If all the remaining passengers who have booked tickets arrive, how many would have to volunteer to take later flights before the Lees would be able to get on the plane?

Number Lines Use the number lines below to solve the problems on page 24. Remember, your number lines do not need to include zero.

1. What is the problem asking for? ______________________________

____

Answer: __________

2. What is the problem asking for? ______________________________

____

Answer: __________

Geometric Drawings

Sometimes it is hard to see how the parts of a geometry problem relate to one another. If you change the height of a cylinder, how does it change the cylinder's volume? How does the radius of a circle affect the circumference? It can also be hard to remember all the different formulas used in geometry. Geometric drawings can help you figure out problems like these.

A geometric drawing is a representation on paper (or some other surface) of a geometric figure. The geometric drawings we make can never be as perfect as the geometric figures they represent, but as long as they are reasonably accurate, they can help us visualize the figures. In fact, it's often impossible to solve a geometry problem without making a drawing.

Geometric Drawings in Action

Imagine this situation. You are taking a math test, and you are faced with the following problem.

> Find the area of a parallelogram whose base is 10 cm and whose height is 8 cm.

You know there is a formula for the area of a parallelogram, but you've forgotten it. What can you do?

You can make a drawing. To make an accurate drawing, of course, you need to know the meanings of all the words in the problem. In this case, a parallelogram is a quadrilateral whose opposite sides are parallel. The base is one of those sides. The height of a parallelogram is the distance between the base and the opposite side. We usually find this by drawing a perpendicular line from the base to the side opposite the base. With this information, you can make an accurate drawing that can help you solve the problem.

Before we start solving any problem, it's important to be clear about what we are trying to find. In this case, the question asked for the area of a specific parallelogram, but our problem is a bit different. Our problem is finding the formula for the area of a parallelogram. Here's a drawing of the parallelogram described in the problem.

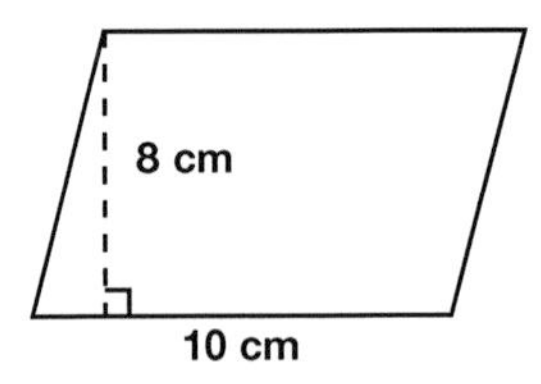

Look at the triangle formed by the left side of the parallelogram, the dotted height line, and part of the base. What if we copied that triangle over to the other side of the parallelogram? We'd have this.

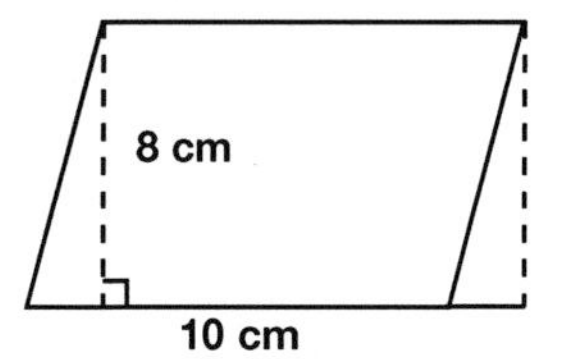

Look at the figure formed by the two dotted lines, the top side of the parallelogram, and the base. What kind of figure is it? It's a rectangle. Its height is 8 cm, and its width is 10 cm. We know that the area of a rectangle is height times width. This means that the area of this figure is 8 cm $\times$ 10 cm or 80 cm^2.

We took a triangle away from one side of the parallelogram and added it to the other side to form a rectangle. These two triangles were congruent—exactly equal in size and shape. This means that the rectangle we formed by moving that triangle has the same area as the original parallelogram. We know that the area of the rectangle is 80 cm^2. So the area of a parallelogram whose base is 10 cm and whose height is 8 cm is 80 cm^2.

Let's review the steps for using a geometric drawing to solve a problem.

1. Make sure you understand all the words in the problem.

2. Determine what you are trying to find. Remember, this isn't always the same as what the question is asking for. Write out what you need to know.

3. Make a drawing using the information in the problem. Make the drawing as accurate as you can. Label the parts of the drawing.

Application Create a geometric drawing to help you solve this problem. Use the grid on page 29 for your drawing. On the lines below the problem, write an explanation of how you approached the drawing, and how you used it to solve the problem. You may find it helpful to include definitions of the terms used in the problem.

The bases (parallel sides) of a trapezoid are 7 cm and 15 cm long. Its height is 5 cm. Find its area.

____

____

____

____

____

____

____

____

____

____

Geometric Drawing Use the grid below to create a geometric drawing for the problem on page 28.

Factor Trees

Many math processes call for finding the factors of numbers—numbers that can be multiplied to produce the original number. Some call for finding the greatest common factor, or GCF, of two numbers. Some call for the prime factors of a number.

There are several ways to find factors. One that can help you visually keep track of all the factors is called a factor tree. This is a diagram with a vaguely treelike shape. It uses "branches" to show the factors of a number.

Before we look any further at factor trees, let's review some terms. You probably know that all whole numbers other than 1 can be written as the product of factors. A **prime number** is a number that has only two factors, itself and 1. An example of a prime number is 13. Its only factors are 13 and 1. A **composite number** is a number that has more than two factors. An example of a composite number is 6. Its factors include 6, 3, 2, and 1. **Prime factors** are factors that are also prime numbers. The **greatest common factor (GCF)** of two numbers is the largest number that is a factor of both numbers.

Factor Trees in Action

Let's make a factor tree for the number 48. To create a factor tree, start by writing the number to be factored on a sheet of paper. The "tree" will spread out, so leave plenty of space on both sides. The larger the number you are factoring, the more space you will need.

Think of two factors of the number. Write the factors below and a little to one side of the original number. Try to keep them on the same level as each other; otherwise your factor tree can get confusing as it goes on. Draw a short diagonal line to connect each factor to the original number.

One easy way to start finding factors is to start with the smallest prime numbers and see if any of them are factors. The smallest primes are 2, 3, 5, and 7. There are some easy tricks to see if 2, 3, or 5 are factors of a number.

- If the last digit of a number is either 0 or an even number, the number is divisible by 2, so 2 is a factor.
- If the sum of the digits of a number is divisible by 3, then the number is divisible by 3, so 3 is a factor.
- If the last digit of a number is either 0 or 5, then the number is divisible by 5, so 5 is a factor.

For example, we can look at the number 30 and say immediately that it is divisible by 2 (last digit is 0), 3 (the sum of the digits is 3, which is divisible by 3), and 5 (the last digit is 0), so all three of these numbers are factors of 30.

Let's try this approach with the number 48. Is 2 a factor of 48? The last digit of the number is 8, which is an even number. This means that 2 is a factor. To find the other factor that, when multiplied by 2, equals 48, divide 48 by 2. $48 \div 2 = 24$. We can write 2 and 24 as factors of 48. Write the factors below the original number. Connect each factor to the original number with short diagonal lines.

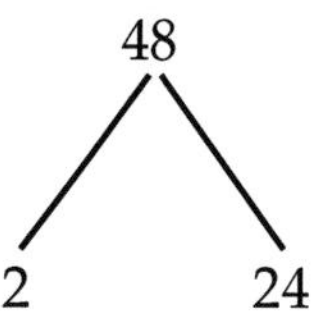

Next, we look to see if these factors can be broken down any further. We know that 2 is a prime factor because it has only 2 factors, 1 and itself. What about 24? Again, let's start by seeing if 24 is divisible by 2. The last digit—4—is an even number, so 24 is divisible by 2. $24 \div 2 = 12$. We can write 2 and 12 as factors of 24.

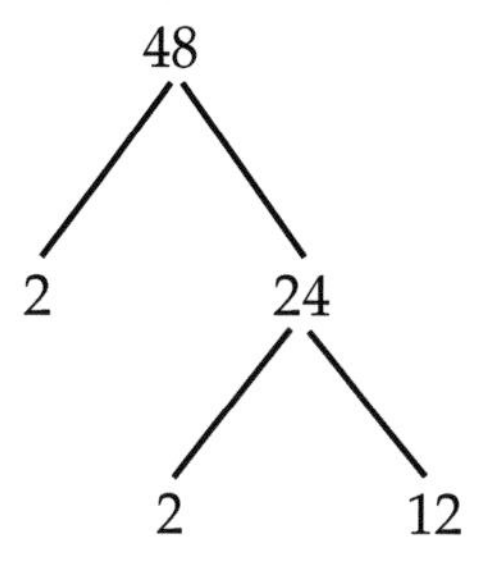

Continue this process, finding factors for each factor, until all your factors are prime numbers.

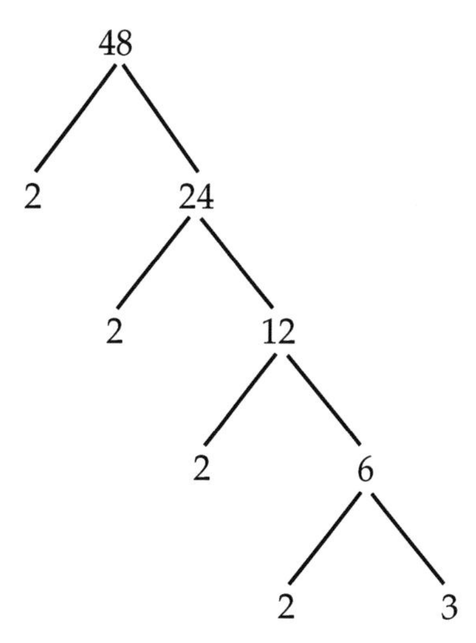

Can any of these factors be broken down further? No, they are all prime numbers. We have factored 48 as far as we can.

The next step is to collect the factors from each step. Any number that is at the end of a "branch" in the "tree" is a prime factor. In this case, the prime factors of 48 are $2 \times 2 \times 2 \times 2 \times 3$.

When we have repeated multiplication of the same factor, it is customary to use exponents as a kind of shorthand. In the case of the prime factors of 48, 2 is a factor four times. We can write this as 2^4. We can then say that 48 is the product of $2^4 \times 3$.

We can use prime factors to find the greatest common factor (GCF) of two numbers. Let's say we need to find the GCF for 48 and 60. Look at the factor trees for both numbers.

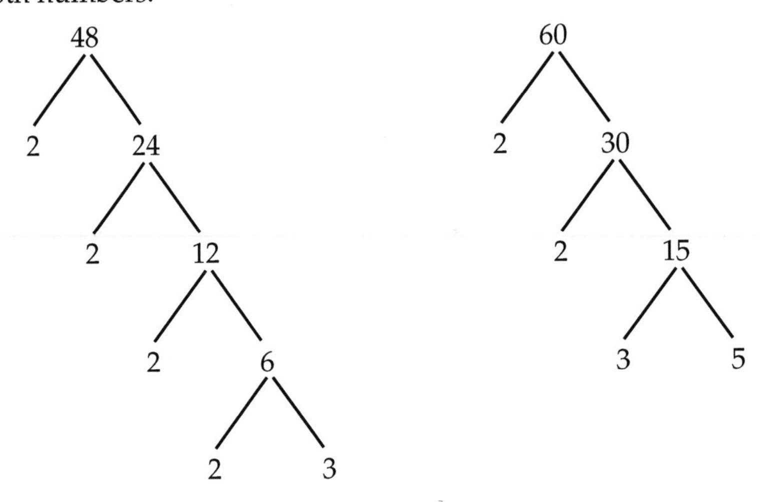

The prime factors of 48 are $2 \times 2 \times 2 \times 2 \times 3$. The prime factors of 60 are $2 \times 2 \times 3 \times 5$. What prime factors do they have in common? Both include two 2s and one 3. This means that the GCF of both numbers is $2 \times 2 \times 3$, or 12.

Let's review the steps in creating a factor tree.

1. Write the number to be factored.
2. See if 2 is a factor of the number. If it is, write 2 a little below and to one side of the number. Connect it to the number with a short diagonal line. Divide the number by 2 to find the other factor that, multiplied by 2, produces the number. Write it below the number in the same way as you wrote 2. If 2 is not a factor, try the other small primes: 3, 5, and 7.
3. Repeat step 2 on the factors you identified in step 2. Write the new factors in the same way, connecting them to the number they factor with short diagonal lines.
4. Continue factoring until all your factors are prime numbers.
5. Collect the factors, and write them using exponential notation to show repeated multiplication of the same factor.

Application Complete the factor trees on page 34 to answer the following questions. Then write the answer to each question on the line.

1. What are the prime factors of 693? Use exponential notation to show repeated multiplication of the same factor.

____

2. What are the prime factors of 1100? Use exponential notation to show repeated multiplication of the same factor.

____

3. What is the greatest common factor (GCF) of 72 and 48?

____

4. What is the greatest common factor (GCF) of 540 and 126?

____

Factor Trees Use these factor trees to answer the questions on page 33. Add or delete lines as needed. Remember, 2 is a factor of a number if the last digit is either 0 or an even number. Three is a factor if the sum of the digits is divisible by 3. Five is a factor if the last digit is either 0 or 5

1. 693

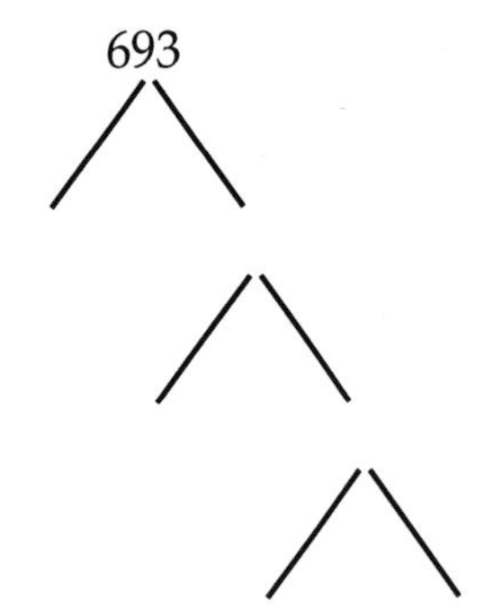

2. 1100

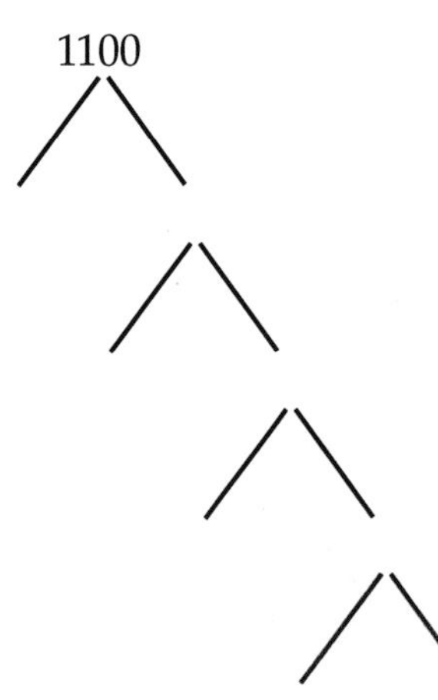

3. 48

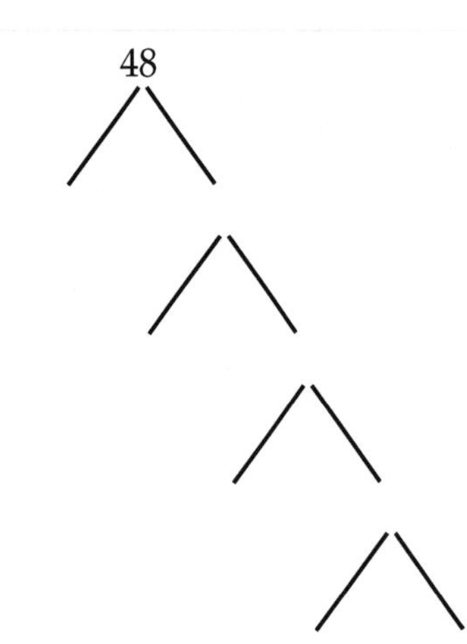

72

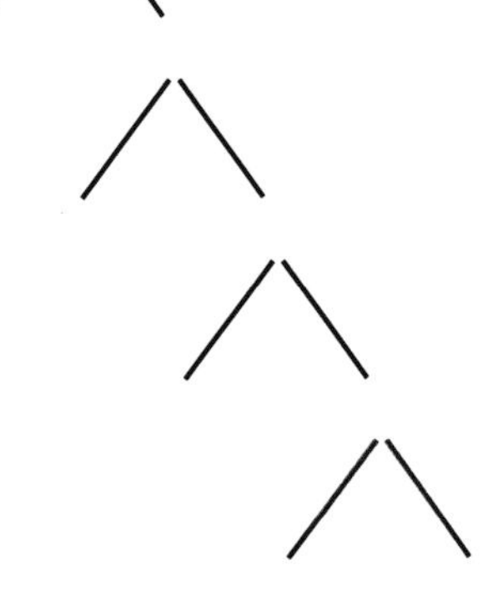

4. 540

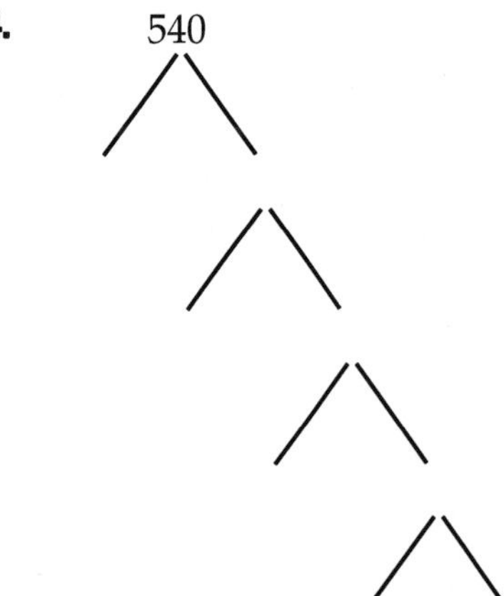

126

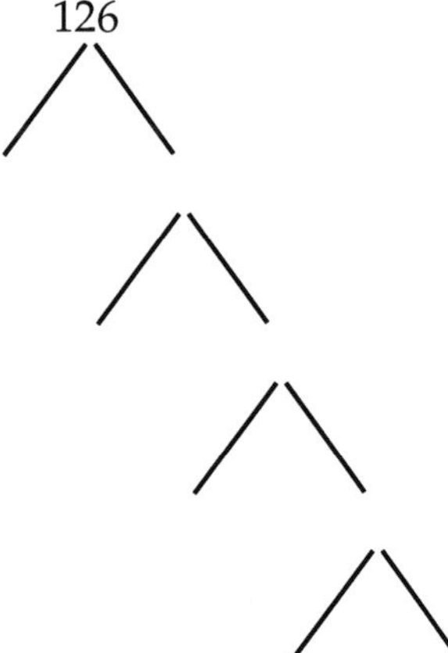

Venn Diagrams

A set is a list of objects in no particular order. Items in a set can be numbers, but they can also be letters or words. Venn diagrams are a visual way of showing how sets of things can include one another, overlap, or be distinct from one another.

Venn diagrams are often used to compare and contrast things. But they are also a useful tool whenever you need to sort and classify information. You can use Venn diagrams to take notes on material that shows relationships between things or ideas. You can also use them to solve certain types of word problems. When a word problem names two or three different categories and asks you to tell how many items fall into each category, a Venn diagram is often a useful problem-solving tool.

Venn Diagrams in Action

A Venn diagram begins with a rectangle representing the universal set. Then each set in the problem is represented by a circle. Circles can be separate, overlapping, or one within another.

When two circles overlap, it means that the two sets intersect. Some members of one set are also members of the other set. For example, let's look at two sets—boys and basketball players. This diagram says that some basketball players, but not all, are boys, and some boys, but not all, are basketball players.

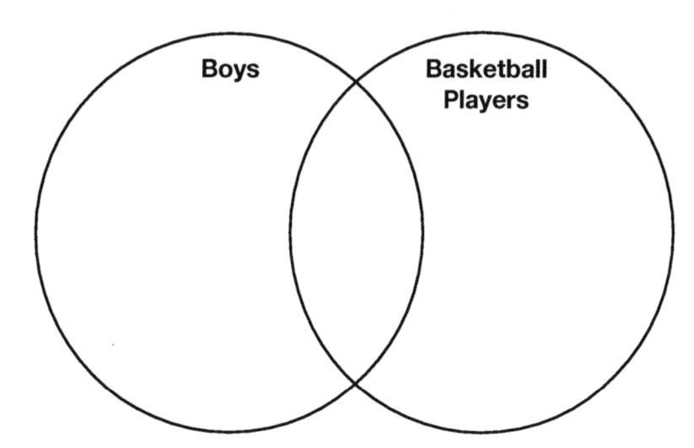

When the circles do not overlap, it means that the two sets do not intersect. No members of either set are also members of the other set. We can say that these sets are mutually exclusive. If something is a member of one set, it cannot possibly be a member of the other set, and vice versa. In this diagram, the two sets are alligators and basketball players. This diagram says that the set of alligators and the set of basketball players do not intersect. If you're an alligator, you don't play basketball, and vice versa.

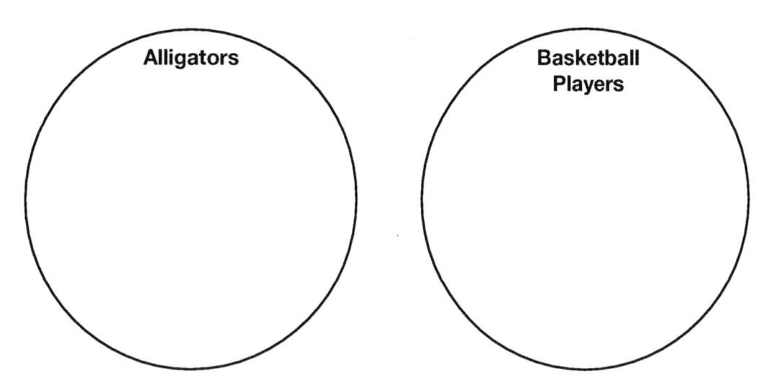

When one circle is completely enclosed within the other, it means that all members of the set in the inner circle are also members of the set in the outer circle. The inner circle is a subset of the outer one. Remember, this does not mean that all members of the set represented by the outer circle are also members of the set represented by the inner circle. This diagram says that the set of alligators is a subset of the set of reptiles. All alligators are reptiles, but not all reptiles are alligators.

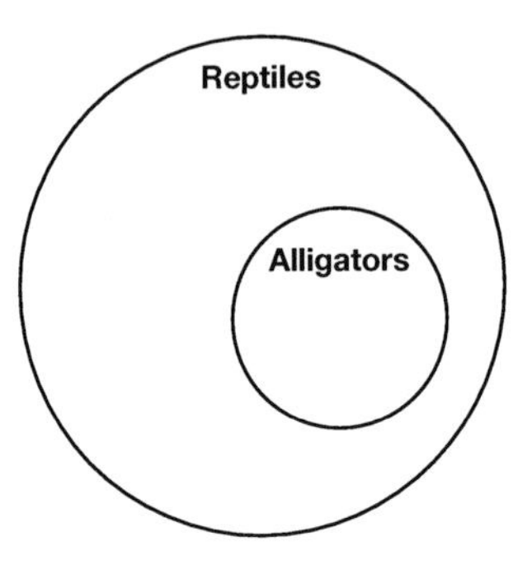

We can use Venn diagrams to keep track of ideas and information as we read. We can also use them to solve word problems. When a word problem gives a few classifications and numbers that fit somehow into the classifications, a Venn diagram can be a useful tool. We can consider each classification as a set and consider the numbers as showing members of that set.

The following word problem is one that could be solved using a Venn diagram. It names several classifications, then asks us to say how many people fit into certain sets and subsets.

> There are 98 students in the ninth grade at Central School. Of those students, 35 play basketball, 25 play soccer, and 9 play both. How many students play neither basketball nor soccer?

The first step in drawing any Venn diagram is identifying the sets involved. What are the sets in this problem? The universal set is all ninth-grade students. Within this, we have other sets: students who play basketball, students who play soccer, students who play both, and students who play neither. Some of these sets are subsets of each other, some intersect, and some are mutually exclusive—members of one cannot be members of another. We are trying to find out how many fit in the last set—students who play neither basketball nor soccer.

Since the set of all ninth-grade students is the universal set and includes all the other sets, we can label the outer rectangle "Ninth-grade students" and write the number 98 in it. All the other sets fit within this set.

How do these sets relate to each other? The sets of students who play basketball and students who play soccer intersect, because we know that 9 students play both. The set of students who play neither does not intersect with the first two sets. We can draw three circles inside the rectangle. Two of the circles will overlap. The third will be completely separate.

Next, we can label each circle and write in the amounts we know for each set. The total for the set of students who play basketball—35—includes the 9 students who also play soccer. The number we write in the outer part of the circle should only show students who play only basketball. We will write 9 in the overlapping area, then write the total number of players minus the overlapping players in the outer area of the circle. 35 – 9 = 26. We will do the same for the soccer players. 25 – 9 = 16. To keep track of the totals, we can write the total for each circle in a box on the margin of the circle.

We can label the third circle, but we don't know what number to put in there. For now, we can just write a question mark, to show that this is what we are trying to find.

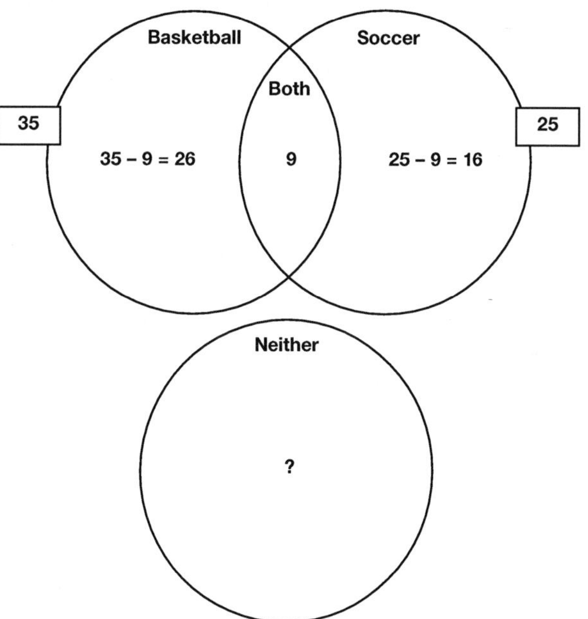

We can now figure out the number that should go in the last circle. This is equal to the total number of students minus the students who play basketball, soccer, or both. We can take the numbers from the Venn diagram and write this statement as an expression, then evaluate the expression. 98 – (26 + 9 + 16) = 47. The number of students who play neither basketball nor soccer is 47.

Let's review the steps in creating a Venn diagram to solve a word problem.

1. Identify the sets involved.
2. Determine how the sets relate to one another. Are they mutually exclusive, intersecting, or subsets?
3. Draw a diagram to show the sets and their relationship to one another. Start with a rectangle for the universal set. Draw separate circles for exclusive sets, overlapping circles for intersecting sets, and circles within circles for subsets. Label each set.
4. Write the information you have been given in each circle. If the total for a set includes subtotals that are also included in another set, subtract the intersecting segment from the number in the main section. To keep track of the totals, write each total in a box on the edge of the circle.
5. Use the diagram to find the missing information.

Application Use the Venn diagram on page 39 to solve the first problem. In the remaining space on page 39, create your own Venn diagram to solve the second problem.

1. A group of 60 students were asked which magazines they read. The survey found that 25 read *¡Mira!*, 26 read *Seventeen*, and 26 read *PC Gamer*. Also, 9 people read both *¡Mira!* and *PC Gamer*, 11 read both *¡Mira!* and *Seventeen*, 8 read both *Seventeen* and *PC Gamer*, and 8 read no magazine at all. How many students read all three magazines? How many students read exactly one magazine?
2. All flooves are shoshniks, but not all shoshniks are flooves. Some rizms are shoshniks but some are not. Rizms that are shoshniks are called vons. Vons are never flooves. Anything that is not a shoshnik or a rizm is a fump. Are all vons shoshniks?

Venn Diagrams

Use the Venn diagram below to solve the first problem on page 38. In the remaining space, create your own Venn diagram to solve the second problem. Write the answer to each problem on the line above the diagram.

1. ______________________________

2. ______________________________

Probability Trees

Thinking about the future can be frustrating. The future doesn't exist. And when it does exist, it won't be the future any more; it will be the present. Trapped as we are at a point in time called "now," the only way we can see into the future is in our minds. When we think ahead, we understand that there is not just one possible future; there are many, some more likely than others. Probability is about expressing mathematically how likely each of a set of possible futures is.

Some probabilities are easy to work out. If a store sells two flavors of ice cream, chocolate and vanilla, it's easy to see the possible outcomes. A customer can buy either a chocolate cone or a vanilla cone. It's also easy to find the probability that a customer will choose vanilla ice cream. There is 1 chance in 2 that it will happen. There is also 1 chance in 2 that the customer will choose chocolate.

But probability also involves events that have many possible outcomes, or that depend on other events happening first—or both. For example, consider a store that sells two types of ice cream, hard and soft. The hard ice cream comes in 4 different flavors: chocolate, vanilla, buttercrunch, and coffee. Soft ice cream comes in 3 flavors: chocolate, vanilla, and chocolate-vanilla swirl. There are 3 different cone types: regular, sugar, and waffle cone. How would you figure out how many possible outcomes there are—how many different types of ice-cream cone a customer could buy? And how would you tell the probability of a customer choosing a hard coffee ice cream in a sugar cone?

When we have probability problems with many possible outcomes, or events that depend on one another, probability trees can help. Probability trees show all the possible outcomes of an event. Whenever a problem calls for figuring out how many possible outcomes there are, and the probability that any one of them will happen, a probability tree can be useful.

Probability Trees in Action

Here's a probability problem. We'll solve it using a probability tree.

> Jay's school wardrobe consists of seven shirts—red, orange, yellow, green, blue, violet, and black—and three pairs of pants—blue, gray, and black. He wears each shirt and each pair of pants equally often. What is the probability that on a given day he will come to school wearing the red shirt and the blue pants?

To draw a probability tree, first identify the starting point. Then identify all the possible choices or outcomes of that starting point. Draw lines to connect the starting point to each possible outcome.

This probability tree will start with the first choice that Jay has to make when he gets dressed in the morning. Will he choose a shirt first? Or will he choose a pair of pants first? We don't know. Does it matter? No. The outcome will be the same either way. For the purpose of the probability tree, let's say he chooses his shirt first. So our probability tree will begin with the point at which Jay chooses his shirt.

The next step is to draw a line from the starting point for each possible choice. Jay has seven shirts to choose from, so we will draw seven lines branching out from the starting point. We also write a label to show which option that line represents.

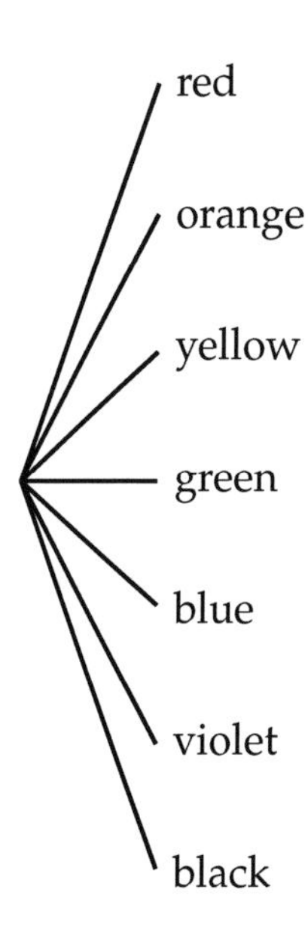

The next step is to identify the next choice to be made, and find all the possible outcomes of that choice for each outcome of the first choice. Again, draw lines to connect each event with its possible outcomes. Keep going, adding lines and possible outcomes, until you've looked at every possibility. When you're done, you'll be able to see just how many outcomes are possible. You'll also be able to see how likely any one outcome is—the probability that that outcome will occur.

Once Jay has chosen a shirt, he has three pairs of pants to choose from. But he has the same three pants choices for each shirt choice. We show this by drawing a line from each shirt choice for each possible pants choice. In this case, we will draw three branch lines out from each shirt choice. Again, we label each line to show which option it represents.

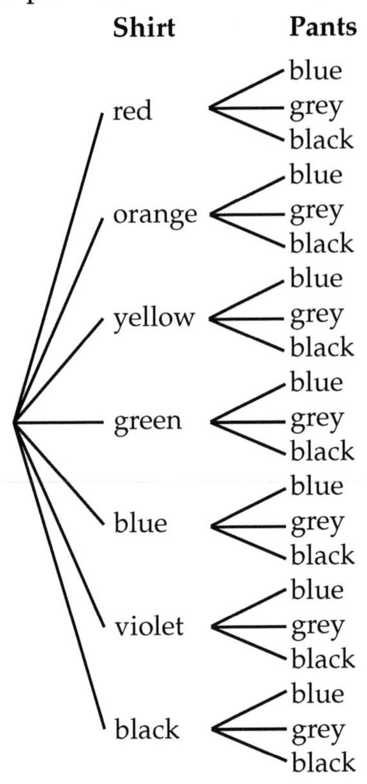

The completed tree shows all the possible outcomes for the available options. In this case, the probability tree shows that Jay has 21 possible shirt-and-pants combinations.

The problem says that each choice of shirt or pants is equally likely. Therefore, the probability that he will wear any one of them—say, the red shirt and the blue pants—is 1/21.

Let's review the steps for creating a probability tree.

1. Identify the starting point.

2. Identify all the possible outcomes of that starting point. Draw lines to connect the starting point to each possible outcome. Label all the outcomes.

3. Identify the next set of possible choices, and find all the possible outcomes of that choice for each outcome of the first choice. Again, draw and label lines to connect each event with its possible outcomes.

4. Continue adding lines and possible outcomes until you have addressed every possibility.

Application

Use the probability trees on page 44 to work out the possible outcomes for each problem below. Then answer the questions that follow each problem.

1. You flip a penny, then a nickel, then a dime, then a quarter. Possible outcomes for each event are heads or tails.

   a. How many possible outcomes are there? _______

   b. What is the probability that all four coins will land heads-up? _______

   c. What is the probability that all four coins will land tails-up? _______

2. Jay has invited 6 friends over to play computer games. Jay has reached Level 7 on the game they're going to play. Of the other players, 2 are also at Level 7. The other 4 have reached Level 8. Each guest arrives separately and rings the doorbell.

   a. How many possible outcomes are there each time the doorbell rings? _______

   b. What is the probability that the first guest to arrive will have reached Level 8? _______

   c. What is the probability that the last guest to arrive will only have reached Level 7? _______

Probability Trees

Complete these partial probability trees to diagram the problems on page 43 Add or delete lines as needed. Then use the completed trees to answer the questions that accompany each problem.

1. Penny Nickel Dime Quarter

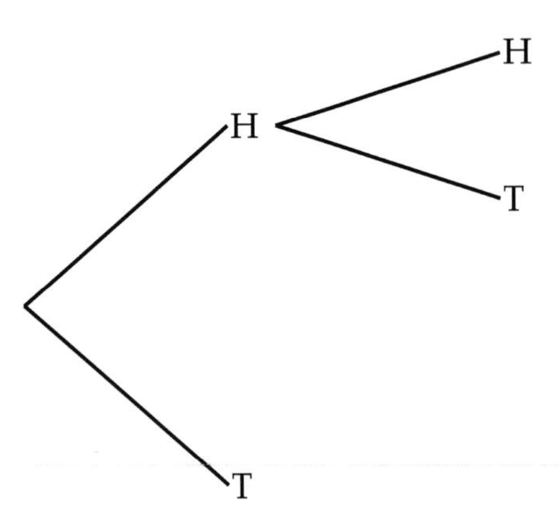

2.

Guest 1

Level 7

Level 8

Attribute Tables

Without logic—an agreed-upon set of rules of reasoning—it would be impossible to clearly define the fundamentals of mathematics. That is why most math programs include logical reasoning and some logic problems.

A logic problem is a word problem that states certain premises and asks you to draw logical conclusions from them. Most logic problems have several subjects, usually people or things. They also list various attributes that belong to the subjects. The problems include clues. These often take the form of negative statements—statements that say that a certain subject does not have a certain attribute. To solve the problem, you have to use the clues to figure out which attributes belong to which subjects. The attributes in most logic problems are exclusive; each attribute can only belong to one subject. If you can figure out that a certain attribute can't possibly belong to two of the three subjects, then you know that it must belong to the third.

To solve logic problems, you need a way to keep track of the subjects and which attributes they have or don't have. A graphic organizer called an attribute table can help. This is a table with a row for each subject in the problem and a column for each attribute. The rows and columns meet to form cells. Because the attributes in logic problems are usually exclusive, you can use **X**s and checkmarks (✔) to show which attribute belongs to which subject.

Attribute Tables in Action

Here's an example of a logic problem. Let's create an attribute table to solve it.

> A doctor, a lawyer, and a minister are friends. Their names are Aiko, Basil, and Camilla. Aiko is not a lawyer. Basil is not 35 years old, but one of his other two friends is. Camilla is not a doctor. The lawyer is not 40, but one of the lawyer's friends is. Camilla is 45. Basil is a minister. What is each person's age and occupation?

To make an attribute table for a problem like this, we start by reading the problem carefully. We need to find the subjects of the problem, their attributes, what values the attributes can have, and whether or not the attributes are exclusive—that is, whether or not each attribute can only apply to one person. In this case, the subjects are three friends: Aiko, Basil, and Camilla. The attributes are age and occupation. The possible values for age are 35, 40, and 45. The possible occupations are doctor, lawyer, and minister.

Once we know the subject of the problem and the attributes, we can set up a table. Draw a grid with one row for each item in the problem and one column for each value of the attributes. This will create a grid with a space, called a cell, where a row and column intersect. Write the information—the problem subjects and attributes—in the grid.

The attribute table for the three friends' problem on page 45 will have a row for each friend and a column for each value of each attribute. One name will be written at the start of each row. An occupation will be written at the top of each of the first three columns, and an age at the top of the next three.

	Doctor	Lawyer	Minister	35	40	45
Aiko						
Basil						
Camilla						

Now that we have the grid for the table, we can look for clues in the problem and use them to fill in the cells. When we know that one of the subjects does not have one of the attributes, we can put an **X** in the cell where that subject's row and that attribute's column meet. When we know that an attribute belongs to a subject, we can put a checkmark (✔) in that cell.

Because the problem says that each attribute value is exclusive, we can make the rule for the table that an attribute column can have only one checkmark in it. Once we put a checkmark in one cell of a column, we can fill in the rest of that column with **X**s. Also, because each person in the problem can have only one occupation and one age, there can be only one occupation checkmark and one age checkmark for each person. So once we put a checkmark on one age or occupation in a row, we can put an **X** in the cells for the other ages or occupations for that person.

Now we can go through the statements in the problem and put the information they contain into the table.

"Aiko is not a lawyer. Basil is not 35 years old, but one of his other two friends is. Camilla is not a doctor." We can put an **X** in the cell where Aiko's row meets the Lawyer column. We can also put an **X** in the cell where Basil's row meets the 35 column, and where Camilla's row meets the Doctor column.

	Doctor	Lawyer	Minister	35	40	45
Aiko		**X**				
Basil				**X**		
Camilla	**X**					

"The lawyer is not 40, but one of the lawyer's friends is." We don't know which person is the lawyer, so we can't put that into the table.

"Camilla is 45. Basil is a minister." We can put a checkmark where Camilla's row meets the 45 column. We can also put a checkmark in the cell where Basil's row meets the Minister column. And because the attributes are exclusive, we can put an **X** in the other cells in the 45 and Minister columns, in the other cells for Basil's occupation, and in the other cells for Camilla's age.

	Doctor	Lawyer	Minister	35	40	45
Aiko		X	X			X
Basil	X	X	✔	X		X
Camilla	X		X	X	X	✔

Now there are only two occupation cells left. Both of them must have checkmarks, because the other cells in the the columns contain **X**s. Since we know that Aiko is neither the lawyer nor the minister, she must be the doctor. And since Camilla is neither the doctor nor the minister, she must be the lawyer.

	Doctor	Lawyer	Minister	35	40	45
Aiko	✔	X	X			X
Basil	X	X	✔	X		X
Camilla	X	✔	X	X	X	✔

We can see from the table that Basil is neither 35 nor 45, so he must be 40. We can put a checkmark in the cell where Basil's row meets the 40 column, and an **X** in the other cell in the 40 column.

	Doctor	Lawyer	Minister	35	40	45
Aiko	✔	X	X		X	X
Basil	X	X	✔	X	✔	X
Camilla	X	✔	X	X	X	✔

This leaves just one blank cell, where Aiko's row meets the 35 column. This means that Aiko must be 35; we can put a checkmark in this last cell.

	Doctor	Lawyer	Minister	35	40	45
Aiko	✔	X	X	✔	X	X
Basil	X	X	✔	X	✔	X
Camilla	X	✔	X	X	X	✔

Now that the table is completely filled in, we can read the solution to the problem. Aiko is the doctor and 35, Basil is the minister and 40, and Camilla is the lawyer and 45.

Let's review the steps for creating an attribute table to solve a logic problem.

1. Identify the subjects and attributes in the problem.

2. Draw a grid with a row for each subject and a column for each attribute value. Write the information from the problem in the grid.

3. Use the clues in the problem to determine which attribute belongs to each subject. Use an **X** to show that a subject does not have a certain attribute. Use a checkmark (✔) to show that a subject does have an attribute.

Application Use the attribute tables on page 50 to solve the following logic problems.

1. Nicky, Anna, and Kit are all athletes. Each plays only one sport. One plays soccer, one plays hockey, and the third one skis. Anna isn't on a team. Nicky, who doesn't like cold weather, is an only child. The hockey player, who is the youngest in her family, likes to spend time with Anna.

   Which sport does each person play?

2. Four couples ate out last week, each on a different night. Ardith went out on Thursday, but not with Alan. Kevin's night out was on Friday. Alan ate out on Sunday, but not with Iris. Jordan and Jen went to the restaurant the day after Iris.

   Which woman went out with which man? And on which night did each couple go out to eat?

Attribute Tables Use the attribute tables below to solve the problems on page 49. Then write the answer to each problem on the lines below the table.

1.

	soccer	hockey	skiing
Nicky			
Anna			
Kit			

____

____

2.

	Tyler	Alan	Kevin	Jordan	Thursday	Friday	Saturday	Sunday
Ardith								
Mari								
Iris								
Jen								
Thursday								
Friday								
Saturday								
Sunday								

____

____

Cause and Effect Map

In this section, we looked at six organizers you can use for solving problems, but there are lots of other ways to approach problems. Here is an organizer you could use in problems that involve causes and effects.

Write each cause in the oval. Write all its effects in the box. Add or delete ovals and boxes as needed.

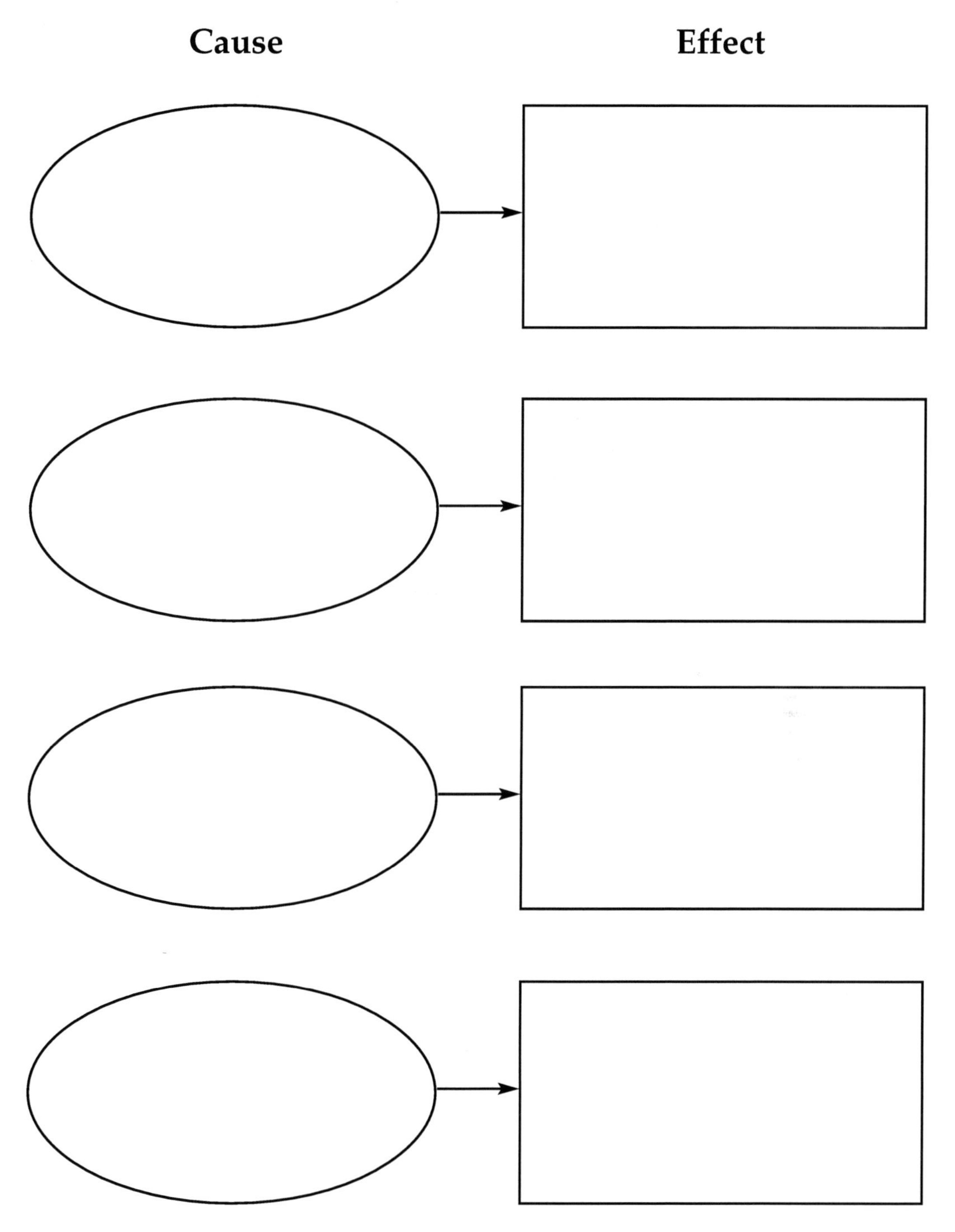

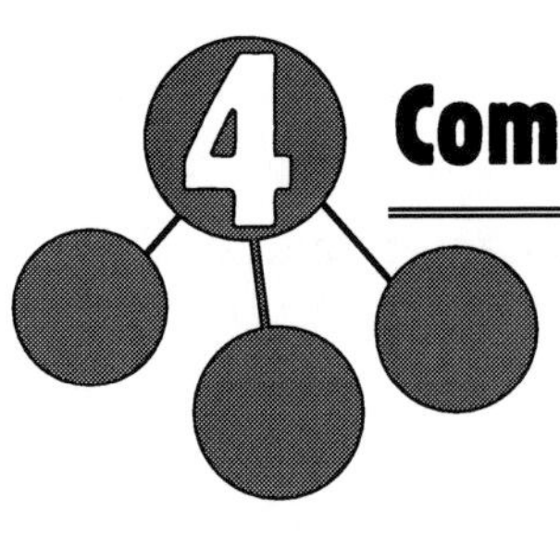

Communicating Mathematical Information

Today's mathematics curriculum recognizes that being able to solve math problems just isn't enough. Both in school and outside school, you will be asked to communicate mathematical information to others.

We have seen that graphic organizers can help you keep track of math concepts or solve math problems. Graphic organizers can also help you communicate mathematical information to someone else. Text is one-dimensional, one word after another. A graphic organizer is two-dimensional. Readers can scan it to find what they are looking for.

This lesson presents four graphic organizers that are often used to present mathematical information. They make it easy for others to understand data that you present and to make comparisons, see trends, and find information.

You can use software, such as Excel, to generate charts like some of the ones presented here. But using software to make charts is like using a calculator to do arithmetic. It can be a great time-saver, but if you don't understand how to do it yourself, you may miss errors. It's important to understand the principles behind each of these charts before you use software to streamline the process. Once you know just how the process works, you can use shortcuts and still tell at a glance if the chart shows the data correctly.

For most sets of information, you could choose one of several different charts to present the information. Look carefully at the information, and think about what you want readers to focus on. You will probably find that one type of chart will be the best for your specific purpose.

Line Graphs

How tall are you? How tall were you three years ago? What about five years ago, or seven, or ten? Your height has probably changed several times in the last ten years. If you wanted to show your height for the past ten years in a graphic way, how would you do it? One approach would be to display it in a line graph. This is a graph that is often used to show how things change over time. Line graphs clearly show trends in data. They can let you make predictions about future trends, too.

Line graphs use two number lines, a horizontal one and a vertical one. The horizontal number line is called the x-axis. The vertical one is called the y-axis. The x-axis often shows the passage of time. The y-axis often shows a quantity of some kind, such as height, speed, cost, and so forth.

Line Graphs in Action

Let's look at a situation that involves change over time and use it to create a line graph. Ashon Johnson's doctor told him that he needed to lose weight. A man of his age and height, the doctor said, should weigh no more than 225 pounds. So, starting on New Year's Day, Ashon went on a diet. He weighed himself weekly and kept records. Here are his records in table form.

Date	Weight (lbs)
1/1	244
1/8	237
1/15	239
1/22	238
1/29	236
2/5	235
2/12	235
2/19	234
2/26	233
3/5	231
3/12	230
3/19	228
3/26	227

Date	Weight (lbs)
4/2	226
4/9	224
4/16	224
4/23	226
4/30	226
5/7	228
5/14	226
5/21	226
5/28	224
6/4	224
6/11	225
6/18	225
6/25	226

Date	Weight (lbs)
7/2	226
7/9	231
7/16	231
7/23	231
7/30	231
8/6	230
8/13	228
8/20	226
8/27	225
9/3	224
9/10	225
9/17	226
9/24	226

Date	Weight (lbs)
10/1	226
10/8	227
10/15	227
10/22	227
10/29	225
11/5	226
11/12	226
11/19	227
11/26	227
12/3	233
12/10	233
12/17	232
12/24	231
12/31	236

After the first year, Ashon went to his doctor for a follow-up visit. To make his progress easier to see, Ashon made a line graph from his data. In making a line graph, it is essential to keep the scales on the axes consistent. Otherwise, the graph can distort the data. To make sure his scale was consistent, Ashon used grid paper to create his graph.

The first step in creating a line graph is deciding what each axis should show. If one of the elements to be shown on the graph is time, this is usually put on the *x*-axis. Ashon's graph was going to show change over time, so he used this as his *x*-axis. He labeled the axis "Dates" and marked it off in one-month increments. The *y*-axis shows the other variable—the thing that changes. In this case, the thing that changed was Ashon's weight. He labeled the *y*-axis "Weight" and marked it off in five-pound increments.

The next step in creating a line graph is to plot points on the graph. For each point, we need to know two pieces of information. We need to know where the point would fall on the *x*-axis and where it would fall on the *y*-axis. These are the coordinates of the point. Ashon's table shows a list of date-weight pairs. For each date there is a weight, and vice versa. In Ashon's graph, the *x*-axis shows time and the *y*-axis shows weight. So the date is the *x*-coordinate in each pair and the weight is the *y*-coordinate. Ashon plotted a point on the graph for each date-weight pair. For example, according to the table, his weight on 1/1 was 244 pounds. Ashon found 1/1 on the *x*-axis, then moved up to 244 on the *y*-axis. He marked a point on the graph to represent that date-weight pair. He did the same thing for all the other pairs in the table.

Once all the points on the graph have been plotted, draw a line to connect them. Here is Ashon's completed graph. All the points have been plotted and connected by a line. He labeled the *x*-axis with every fifth date, starting with the first. The graph gives a clear picture of how Ashon's weight varied over the course of the year.

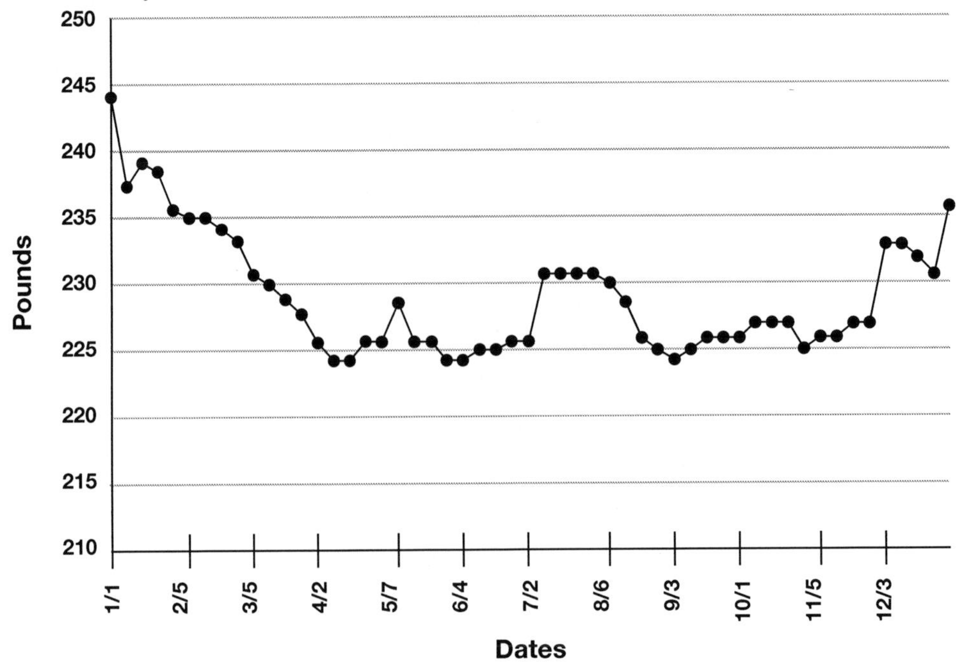

Ashon brought the graph to his doctor. The graph showed them a pattern that would have been harder to see in the table. They saw that Ashon had been at or near his goal weight of 225 for most of the year. They also saw that he had gained weight during two periods: July and December. "Of course!" said Ashon. "There were a lot of holiday parties then. I guess I got carried away." Having seen the graph, Ashon knew what he had to do to keep his weight down and stay healthy.

Let's review the steps for creating a line graph.

1. Decide what each axis should show. Label both axes. Mark appropriate increments on each axis.

2. Plot points on the graph, using the coordinates for each point.

3. Join the points with a line.

Application Use the grid on page 57 to make a line graph of the data in this problem. Then answer the questions below the graph.

Alani drove to a friend's house about 14 miles away. To get there she drove through local streets near her own house, got onto the Interstate for a few miles, got off, and drove through local streets to her friend's house. The trip took 23 minutes.

Alani's mother is an automotive engineer. She had fitted the car with the prototype for a new event-data recorder. The recorder tracks time and distance traveled, then prints it out. At the end of Alani's trip, the device printed out the following data.

Time (hh:mm:ss)	Distance (miles)
0:00:00	
0:01:00	0.33
0:02:00	0.63
0:03:00	0.98
0:04:00	1.32
0:05:00	1.63
0:06:00	2.22
0:07:00	2.97
0:08:00	3.73
0:09:00	4.43
0:10:00	5.20
0:11:00	5.88
0:12:00	6.60
0:13:00	7.40
0:14:00	8.28
0:15:00	9.20
0:16:00	10.10
0:17:00	11.03
0:18:00	11.90
0:19:00	12.32
0:20:00	12.70
0:21:00	13.10
0:22:00	13.47
0:23:00	13.63

Line Graph Use the graph below to show the data on page 56. Then answer the questions below the graph.

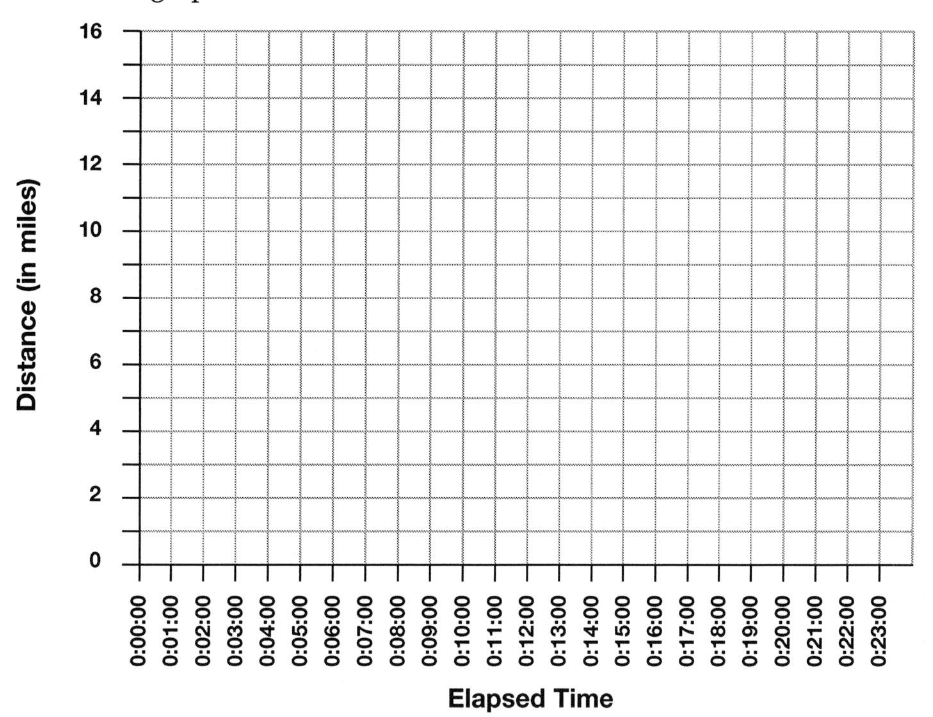

1. Based on the graph, at what time (in minutes) during the trip did Alani get onto the Interstate? ________

2. When did she get off? ________

3. How many miles did she travel on the Interstate? ________

Bar Charts

You have learned that tables are a good way to record and store information. But they aren't always the best way to present information to others. For example, imagine that you and three friends were playing fantasy baseball. You might use a table to store information about each team's wins and losses.

Team	Wins	Losses
Carl's Crazies	92	70
Dee's Dominators	79	83
Johnstown Wonders	71	91
Pine Tree Winners	66	96

The table gives all the information. But it doesn't really show how the four teams compare to each other. Here is the information presented in a different way—using a bar chart.

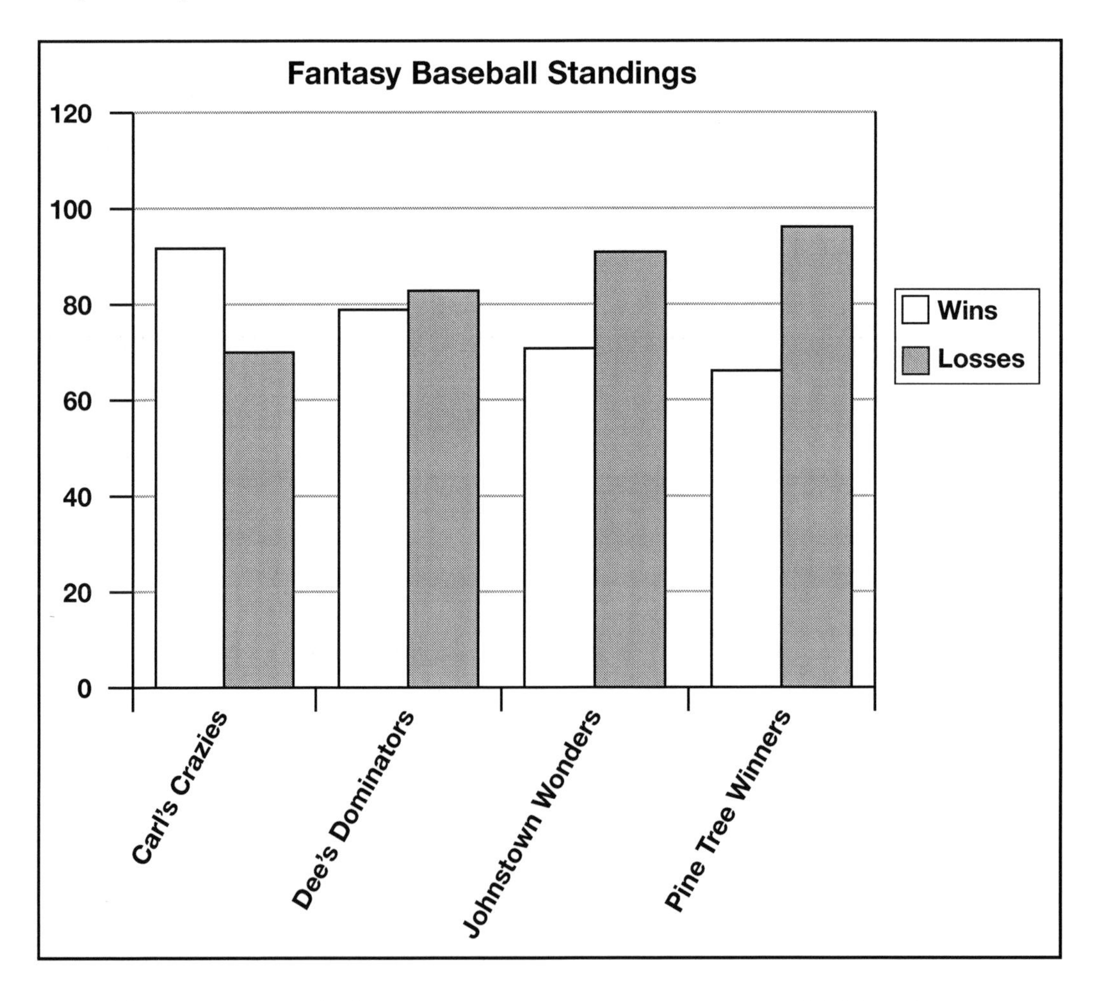

Now it's easy to see which team has won most games and which has lost most.

Bar charts are useful when you want to compare things or to show how one thing changes over time. They are a good way to show overall trends. Bar charts use horizontal or vertical bars to represent data. Longer bars represent higher values. Different colors can be used to show different variables. When you look at a bar chart, it's easy to see which element has the greatest value: It's the one with the longest bar.

Bar charts have an *x*-axis (horizontal) and a *y*-axis (vertical). If the graph is being used to show how something changes over time, the *x*-axis has numbers for the time period. If the graph is being used to compare things, the *x*-axis shows what things are being compared. The *y*-axis has numbers that show how much of each thing there is.

Bar Charts in Action

Let's look more closely at that fantasy baseball league. Here are batting statistics for some of the players in the league. We can use the information in the table to make a bar chart that compares these players.

Name	AB	R	H	2B	3B	HR	RBI
Joe Crede	318	45	76	19	0	13	45
Johnny Damon	392	80	121	26	2	13	53
Aubrey Huff	374	55	104	13	1	17	62
Derek Jeter	392	59	107	24	0	14	49
Jacques Jones	349	44	87	15	1	16	54
Victor Martinez	334	57	103	26	1	17	78
Brian Roberts	387	63	102	30	1	2	30
Iván Rodríguez	352	49	125	24	1	13	63
Gary Sheffield	353	72	105	17	0	20	72

KEY

- **AB** at-bats; the number of times a batter has come up to bat
- **R** the number of times a batter has crossed home plate
- **H** hits that resulted in the batter safely reaching a base
- **2B** hits that resulted in the batter reaching second base
- **3B** hits that resulted in the batter reaching third base
- **HR** hits that resulted in a home run
- **RBI** runs batted in; the number of times a batter has made it possible for teammates to score

To create a bar graph, start by deciding what you want to show in the graph. Do you need to show all the information you have, or would your chart be clearer with just some of the information? This table has a lot of information. There are nine players in the table and seven statistics for each player. If we made a separate bar for each statistic and for each player, the chart would have 63 bars. That probably wouldn't be very easy to read. We could make a chart for each player that shows only that player's statistics. Or we could choose the two or three statistics that give the most information, then make a chart that just shows those statistics. Since we want to compare the players, it would be most helpful to have all the players on one chart. We will take the second option: a chart that shows just a few statistics for all nine players. Let's take at-bats, hits, and runs batted in.

Before we start making the chart, we need to decide on the overall layout. Bar charts can be horizontal, with bars that go from one side to the other, or vertical, with bars that go up from the bottom. If the labels on the bars are very long, you may want to make the chart horizontal. That way you have more space in which to write. For this chart, we will use vertical bars. The x-axis will show the players and statistics. The y-axis will show quantity—how many of each statistic each player had.

We know what information we will put in the chart and what each axis will represent. Now we need to decide how to organize the information on the x-axis. We have nine players, and we have three pieces of information for each player. That means we will have 27 bars on the x-axis. We could organize them so that all the at-bats were together, all the hits were together, and all the runs batted in were together. But the purpose of the chart is to compare the players. The comparison will be easier to see if we keep each player's statistics together. We can arrange the bars in groups of three, with a different shading or color for each statistic.

The next step is deciding on the values for the y-axis. The highest value on the y-axis should be just a bit higher than the highest value in the data we are trying to show. In this case, the highest value is 392. We could use a maximum value of 400 on the y-axis. We will also need to show values at intervals along the y-axis. Again, look at the data to see what would make sense. In this case, intervals of 20 would probably work well. Depending on the values on the y-axis, you may want to add a label explaining what this axis shows. For example, if you were comparing climates, intervals might stand for inches or centimeters of rain. Labeling the y-axis "Inches of rain each year" would help readers make sense of the chart.

Now we're ready to create the chart. To make a chart that really shows how values relate to one another, it's important to have all the bars use the same scale. This is easiest if you use grid paper. Mark the y-axis on the chart. Label 0, the maximum value, and the intervals. To see how wide the chart should be, look at the number of bars you will be drawing. We need 3 bars for each of 9 players, for a total of 27 bars. The chart should be at least 27 units wide.

Start with the first bar—in this case, the first player. Joe Crede had 318 at-bats. Find 318 on the y-axis. It will be a little below the 320 mark. Draw a short horizontal line at that point. Next, draw two straight lines up from the x-axis, one to each end of the horizontal line, so that they form a bar. Choose a color for this bar, and color it in. You will use this color for all at-bats.

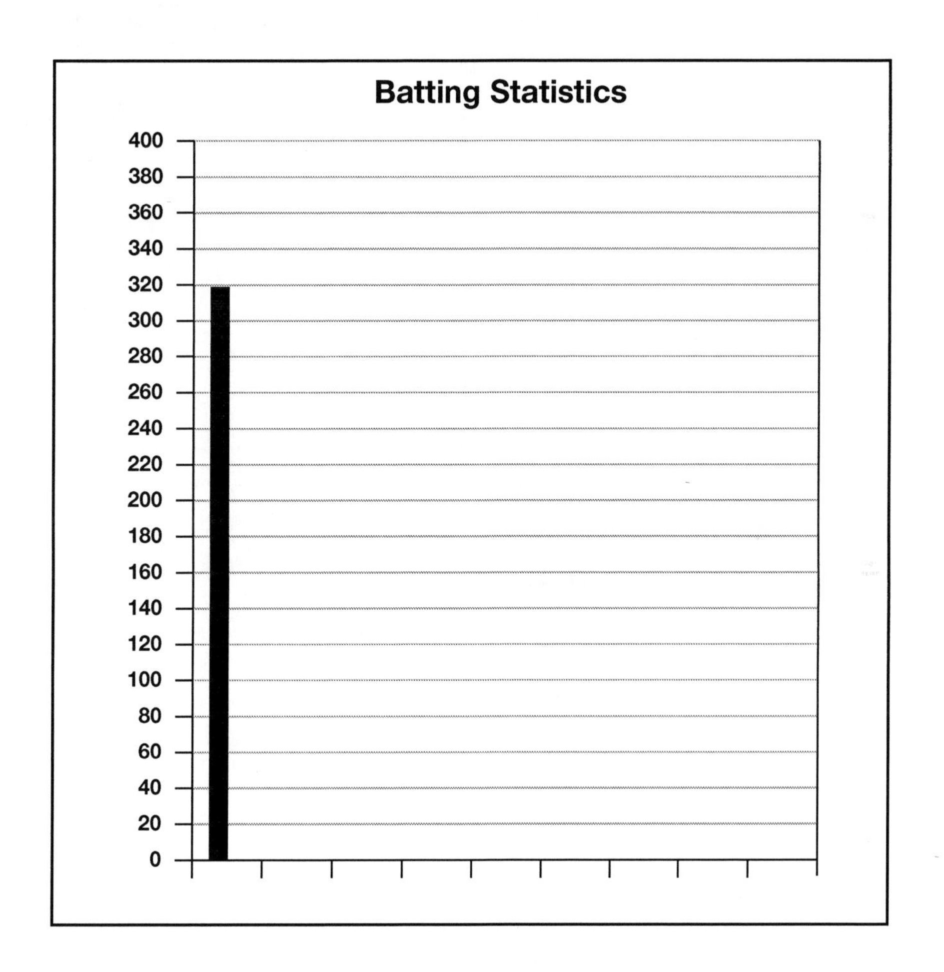

Now do the same thing for Joe Crede's other statistics, using a different color for each bar. Write the player's name under the group of three bars. Draw a small box in one corner, and make a key showing what each color represents.

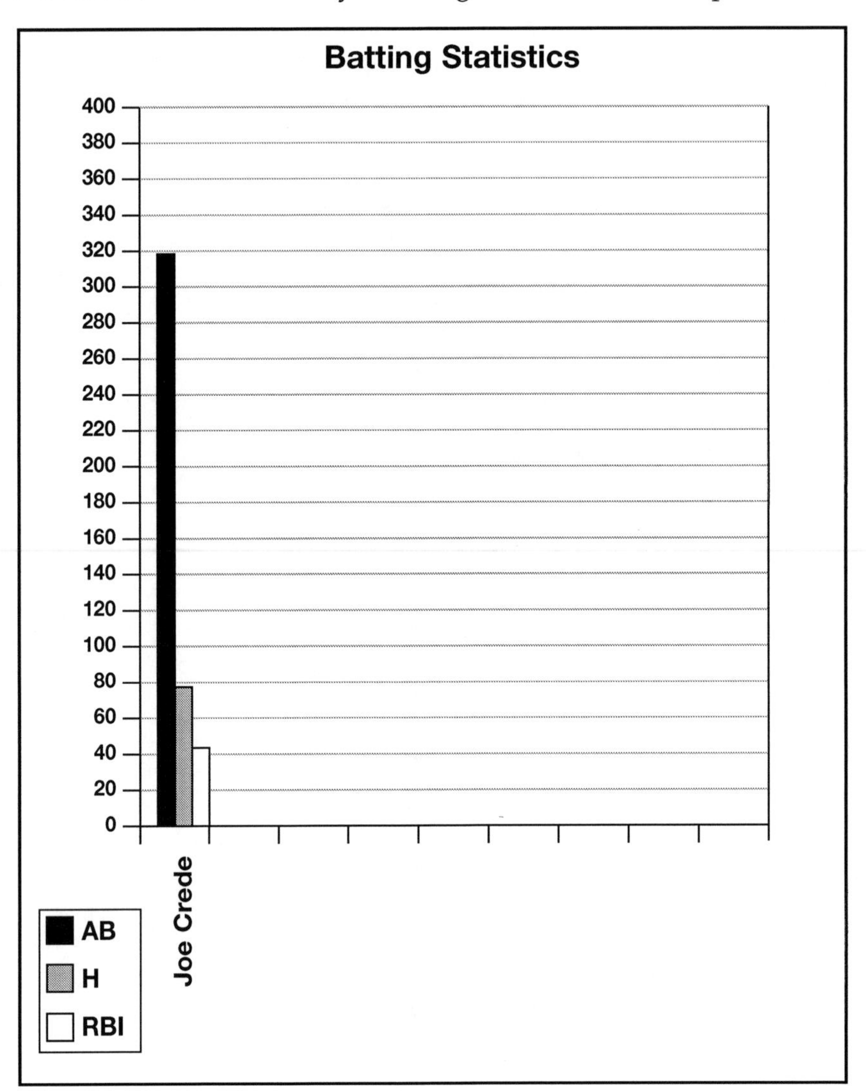

Leave a slight space to the right of the third bar, then make bars for the next player's statistics. Color the bars according to the key. Continue in this way until you have made bars for all the statistics for each player. Finally, add a title for your chart.

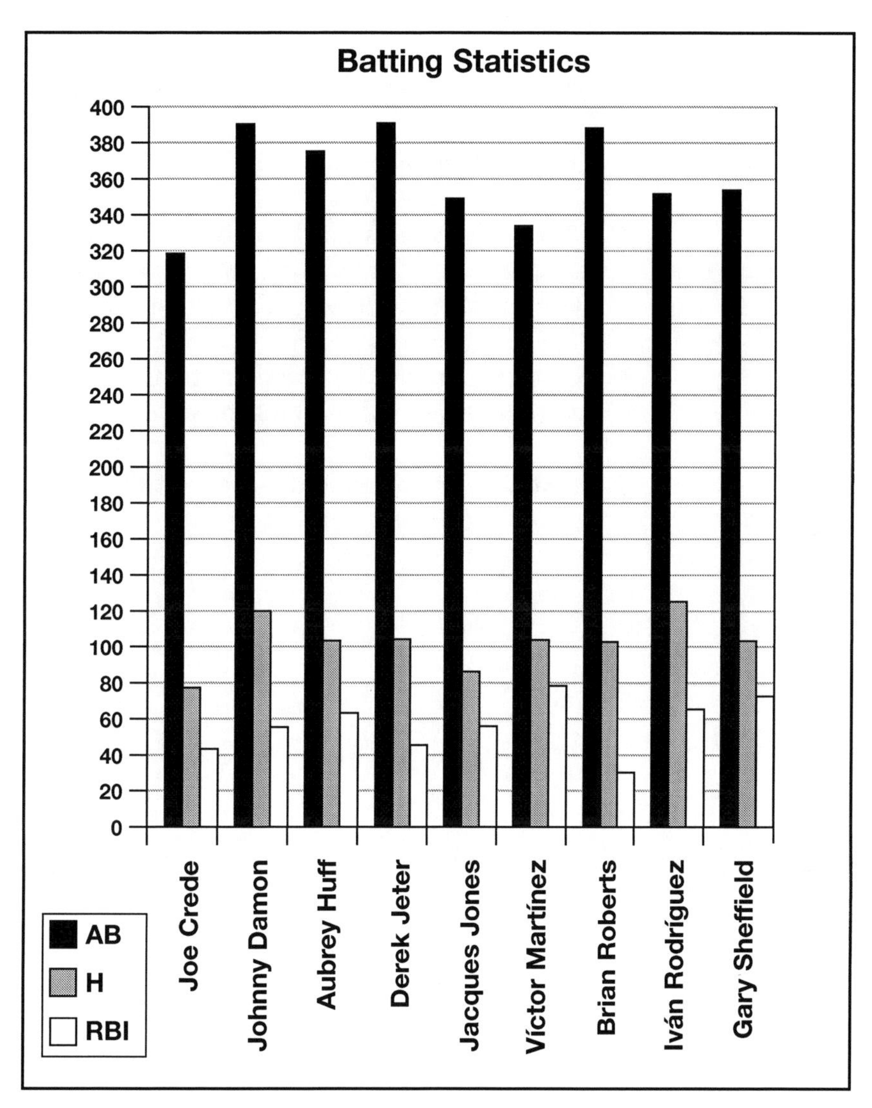

The completed chart makes it easy to compare these players' batting statistics.

Let's review the steps for making a bar chart.

1. Choose the information you want to show.
2. Choose a format for your chart, with either horizontal or vertical bars.
3. Decide what each axis will show.
4. Choose the values for the y-axis. If necessary, give the y-axis a label.
5. Base the width of the chart on the number of bars it will have.
6. For each bar, find its value on the y-axis, then create a bar by drawing lines to connect that value to the x-axis. Label each bar along the x-axis.
7. Create a key, and give the chart a title.

Application Use the blank bar chart on page 66 to show the information in the table below.

Ms. Washington gave a math test with 15 items on it. This table shows how many of the 25 students who took the test got each item correct. Make a bar chart to display the data in this table. Then answer the question below the chart.

Item Number	Problem Type	Number of Students Who Answered Correctly
1	Word problem	24
2	Word problem	23
3	Word problem	24
4	Adding fractions	3
5	Adding fractions	5
6	Adding fractions	2
7	Solving an equation	15
8	Solving an equation	16
9	Solving an equation	17
10	Geometry	22
11	Geometry	25
12	Geometry	23
13	Solving an inequality	20
14	Solving an inequality	19
15	Solving an inequality	20

Bar Chart Use the bar chart below to show the information in the table on page 65.

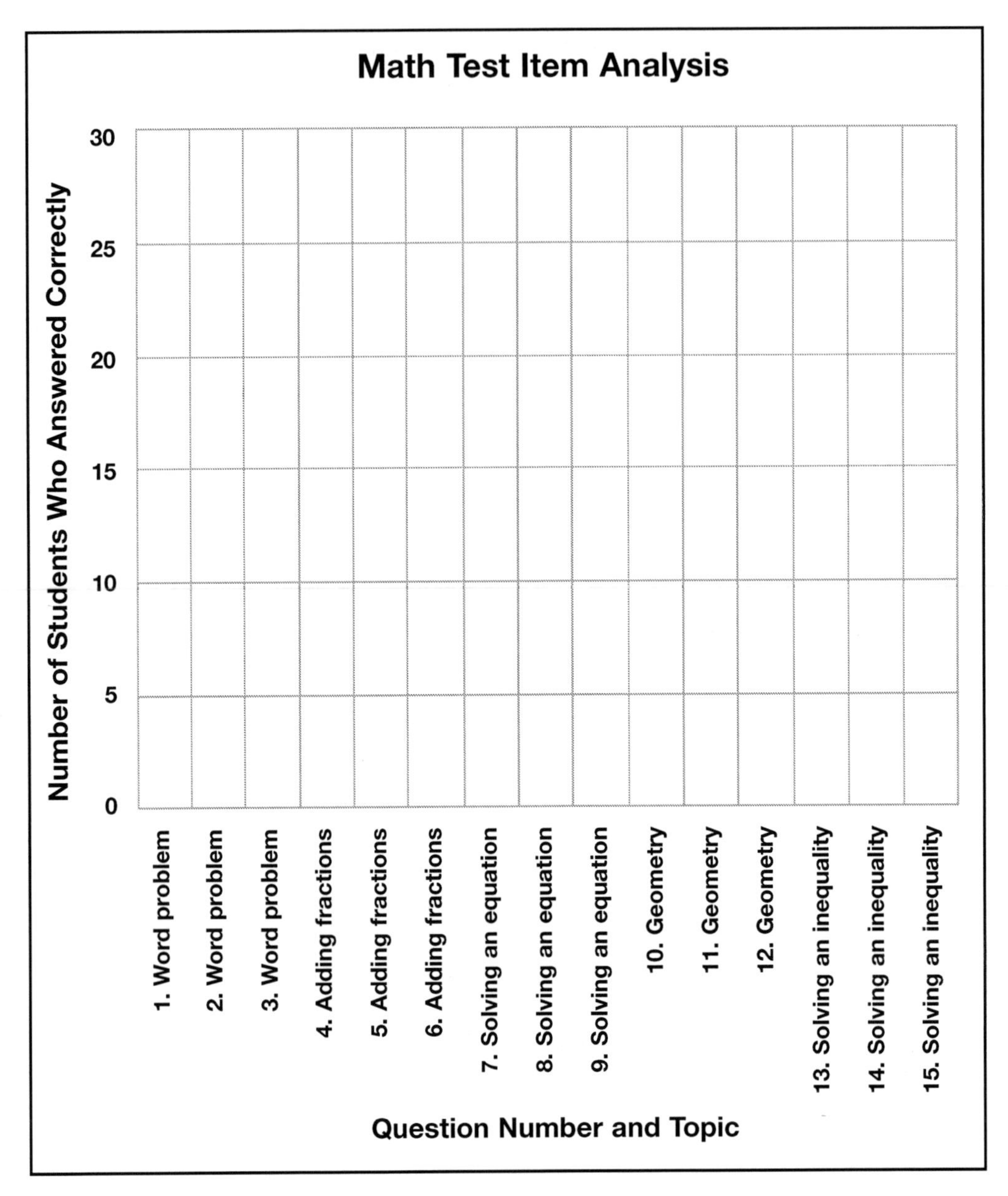

Based on the chart, what topics should Ms. Washington go over again in class?

Pie Charts

Tables are a good way to collect information, and bar charts are a good way to compare different things. What if you wanted to compare parts of one thing? For example, say you spend an average of $100 every month: $40 on meals and snacks, $30 on movies and other entertainment, $20 on clothes, and $10 on other expenses. You're curious about how you're spending the money. You could make a table of all your expenses. You could make a bar graph showing how much you spend in each category. But if you really want to know how your expense categories relate to one another, you might make a pie chart.

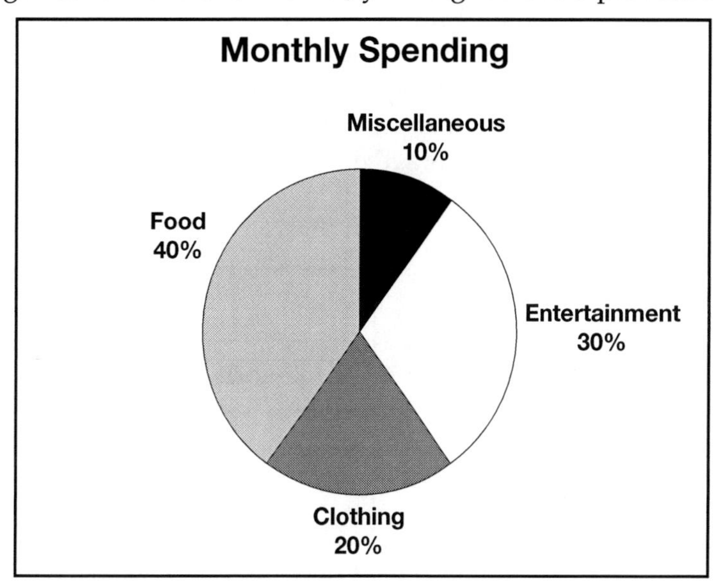

The pie chart shows clearly how the parts relate to the whole. You can see that you spend more on food every month than you spend in any other single category.

Pie charts are a good way to show how the parts of a whole relate to one another. You can use pie charts as we did above, to show categories in a budget. You can use them to show percentages of voters who chose a certain candidate, or survey respondents who chose a certain option. Whenever you are working with information that can be shown as percentages of 100, or parts of a whole, you could consider using a pie chart.

Unlike other charts and graphs, pie charts don't use a set of axes to plot points. Instead, they are based on a circle. Another difference is that pie charts don't work with numbers in the same way as other charts. They show percentages. The whole circle is 100%. Each segment of the circle represents part of that 100%. In the pie chart we made for monthly spending, the percentages happened to match the dollar amounts. This is because the total amount being spent was $100, so each dollar was 1% of the total.

Pie Charts in Action

Imagine that, for a science project, you have been asked to keep track of how you spent your time over the course of five days, then create a chart to show the information—without using software for any part of the project.

You have tracked your time on a data chart that was broken down in 15-minute intervals. For each interval, you wrote down what you were doing. Once you had completed the data chart, you identified some general categories for how you spent your time: school, homework, sleep, television, socializing, and other. ("Other" was a catch-all category for activities that took up time, but not enough time to have a whole segment each. It included activities, such as going to the library, waiting for the school bus, and so forth.) You used the categories to make a table showing how you spent your time over the course of five days, and the average time (in minutes) spent daily on each activity. Here is the table:

Activity	Min/5 days	Min/day
School	1650	330
Homework	710	142
Sleep	2450	490
Television	600	120
Socializing	800	160
Other	990	198
Total	7200	1440

You have decided to present the information in a pie chart. This format will show about how much of each day you spent doing each activity.

Because the project didn't allow you to use any software, you have to create the chart by hand, using a calculator, a compass, and a protractor.

The first thing you need to do is to change the data into percents of the whole. The whole was 1440 (the total number of minutes in 24 hours). Use a calculator to find what percentage of 1440 each value is, then round each one off to the nearest whole percent.

Activity	Min/day	Percent of 1440
School	330	23
Homework	142	10
Sleep	490	34
Television	120	8
Socializing	160	11
Other	198	14

Next, figure out how to fit these percentages into a circle. A circle is divided into 360 degrees. To find how many degrees each percentage should cover, you need a conversion factor. The total number of minutes is 100%. There are 360 degrees in a circle. So to find the number of degrees, you need to divide each percentage by 100, then multiply it by 360. Dividing by 100, then multiplying by 360, is the same as multiplying by 3.6. So you can use this as the conversion factor. To find the number of degrees each segment should cover, multiply each percentage by 3.6. Round off the answers to the nearest whole degree. To make sure the math is correct, add all the degrees in the column; they should total 360, or a whole circle.

Activity	Min/day	Percent of 1440	Number of degrees
School	330	23	83°
Homework	142	10	36°
Sleep	490	34	122°
Television	120	8	29°
Socializing	160	11	40°
Other	198	14	50°

Now you are ready to draw a pie chart. Use a compass to draw a circle. Decide where one segment will start. Draw a line from the center of the circle to the circumference. Place a protractor on the line. The center of the protractor's base at the center of the circle, and the 0 on the protractor scale should be on the line you just drew. Measure the degrees required for the segment, starting from the 0 on the protractor scale. Make a mark on the circumference of the circle. Draw a straight line from the center of the circle to the mark on the edge. For the next segment, use your new line as the base line for the protractor. Keep going until you have drawn all the segments.

Label each segment to show what it represents and how many minutes were spent doing it. Finally, color each segment a different color.

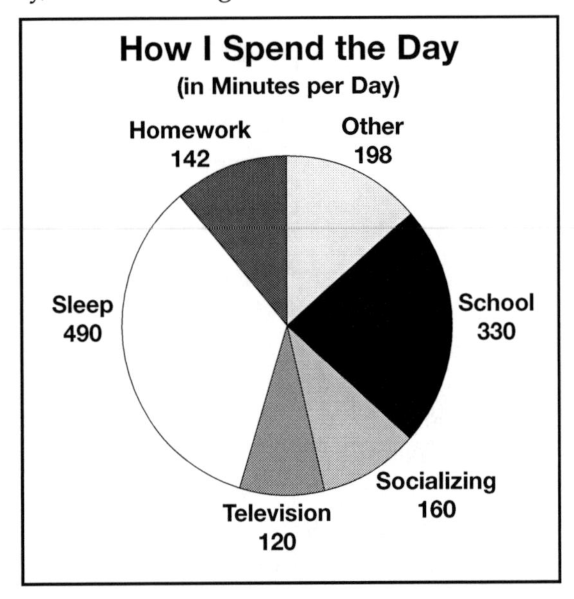

Let's review the steps in creating a pie chart.

1. Gather the data for the chart. Decide on the categories you want to show. Some smaller categories may be best grouped together.

2. Change the data into percents of the whole. Check your math—the percents for all the segments should add up to 100%.

3. Convert percentages into degrees of a circle by multiplying by 3.6. Check your math—the degrees for all the segments should total 360°.

4. Use a compass to draw a circle, then use a protractor to find the number of degrees for each segment.

5. Label and color each segment. Give your chart a title.

Application Use the blank pie chart on page 72 to display the information below.

> The senior class of Padua High recently held an election for class president. There are 137 students in the class. Three students ran for president. Elli Rodriguez, who has been a student at Padua since her sophomore year, received 88 votes. Fela Johnson transferred to Padua in his junior year. He received 28 votes. Honor Pulaski, who started at Padua as a freshman, received 17 votes. There were also 4 votes for non-candidates: 2 for Vince, the school mascot; 1 for the actor Johnny Depp; and 1 for Ms. Morton, the school principal.

Use the space below to work out how many degrees each segment should cover. Then use the blank pie chart on page 72 to show the election results. On the chart, include the percentage of the vote each candidate received.

Pie Chart Use this blank pie chart to display the information on page 71. Each segment of the blank chart equals 18°.

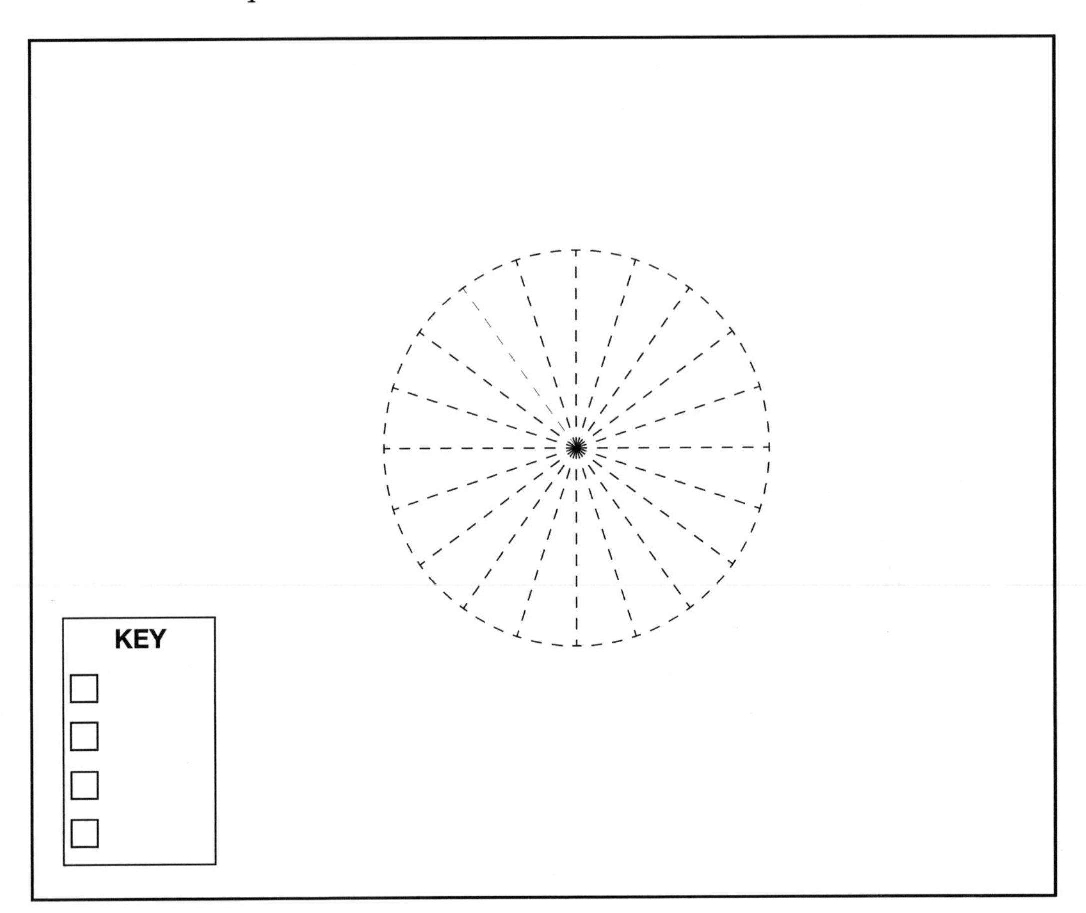

Stem-and-Leaf Plots

Line graphs and bar charts are good ways to show patterns in sets of data. But one problem with these organizers is that they only show the patterns. The actual values that the patterns are based on disappear in the chart. However, there is a way to organize data that shows both the patterns and the values: a stem-and-leaf plot.

A stem-and-leaf plot displays values by breaking each one into a "stem" and a "leaf." The leaf is the digit in the place farthest to the right in the number. The stem is what is left when the leaf is dropped. Essentially, a stem-and-leaf plot arranges the numbers in a group according to place value.

This arrangement makes it easy to see which numbers in a set occur most often. Stem-and-leaf plots are useful for showing how values in a set are distributed. They work well for data sets that include specific values, not ranges. For example, a stem-and-leaf plot would be a good way to show how many books each student in a read-a-thon read. It wouldn't work as well to show how many read fewer than 5 books, how many read between 5 and 10 books, and how many read more than 10 books. Also, stem-and-leaf plots are not a good way to display data sets with a large number of values. If you want to show how often certain values appear in a set with a fairly small number of specific values, a stem-and-leaf plot may be a useful display.

Stem-and-Leaf Plots in Action

We noted that stem-and-leaf plots are useful for sets of data that have specific values. One area in which we often find data like this is in sports. Let's use a cyclist's practice log to make a stem-and-leaf plot.

Jana has been training for a 250-mile fund-raising bike ride, to be held over three days. Here are the distances she rode last month, in miles: 25, 37, 42, 35, 28, 14, 28, 57, 52, 19, 26, 24, 22, 36, 45, 38, 18, 29, 16, 24, 17, 49, 50, 16, 23, 37, 35, 28, 46, 52.

Start by writing the values in order, from least to greatest.

> 14, 16, 16, 17, 18, 19, 22, 23, 24, 24, 25, 26, 28, 28, 28, 29, 35, 35, 36, 37, 37, 38, 42, 45, 46, 49, 50, 52, 52, 57

Next, identify all the possible stems in the group of values. Remember, the leaf is the digit farthest to the right, and the stem is everything else. Since all the numbers in this set are two-digit numbers, the stems will be the numbers in the tens place. The possible stems are 1, 2, 3, 4, and 5.

Now we're ready to plot the data. Start by drawing a vertical line on a sheet of paper. This line will separate the stems and leaves. Write each possible stem on the left of the line.

Stem	Leaves
1	
2	
3	
4	
5	

Next, write each leaf on the right side of the line, opposite its stem. When two or more numbers have the same stem, write the leaves one after the other, starting with the smallest digit. If the same digit appears as a leaf more than once, write it out each time. You don't need any punctuation between the digits.

Stem	Leaves
1	4 6 6 7 8 9
2	2 3 4 4 5 6 8 8 8 9
3	5 5 6 7 7 8
4	2 5 6 9
5	0 2 2 7

The completed plot shows that most of Jana's training rides last month were between 20 and 30 miles. The distance she rode most often—the mode of the data set—was 28 miles.

You can use a back-to-back stem-and-leaf plot to compare two sets of data. Let's say Jana wanted to see how her training rides this month compare to last month's rides. She could draw another line on the other side of her stems, then plot a new set of leaves on the same stems. Here are Jana's distances from this month, arranged from least to greatest.

16, 17, 17, 19, 22, 25, 26, 27, 28, 28, 29, 32, 35, 35, 36, 37, 37, 38, 39, 39, 42, 45, 46, 49, 50, 52, 52, 55, 56, 57

To add these values to the stem-and-leaf plot, draw another vertical line to the left of the stems. Using the same stems, plot the new values to the left of the original plot.

6 7 7 9	1	4 6 6 7 8 9
2 5 6 7 8 8 8 9	2	2 3 4 4 5 6 8 8 8 9
2 5 5 6 7 7 8 9 9	3	5 5 6 7 7 8
2 5 6 9	4	2 5 6 9
0 2 2 5 6 7	5	0 2 2 7

Jana can see that more of this month's rides are between 20 and 30 miles. She has also done more rides of more than 50 miles.

The steps for making a stem-and-leaf plot stay the same, even when the data set includes larger or smaller numbers—even decimals. The leaf is always the digit farthest to the right. The stem is always everything that is left when the leaf is taken away. For example, for the number 492, the leaf is 2 and the stem is 49. For the number 1347, the leaf is 7 and the stem is 134. For the number 19.2, the leaf is 2. The stem is 19. For the number –14, 4 is the leaf, and –1 is the stem. To make a stem-and-leaf plot with single-digit numbers, such as 1 or 5, just use 0 as the stem.

Let's review the steps in making a stem-and-leaf plot.

1. Put the values in order.
2. Write out the stems.
3. Add the leaves.

Application The paragraph below shows a teacher's notes about a recent test. The teacher wants to display the data in a stem-and-leaf plot. Use the lines below the paragraph to list and organize the data. Then use the blank stem-and-leaf plot on page 77 to display the information. When you have completed the plot, look at it carefully. On the lines below the plot, write one or two sentences describing what the plot shows about the results of this test.

> The scores on the test I recently gave to 15 students are mostly good. Some students, however, need improvement. Four students scored in the 80s or above: Kayla, Bree, Aaron, and Myke got 88, 83, 84, and 88, respectively. Kris got a 96, Holly a 97, and Jordan scored a perfect 100. Fara, who has been doing poorly up to now, got a 77, which shows she has been working hard. Four others scored in the 70s: Declan, 76; Josh, 74; Mirka, 74; and Ethan, 70. Sam got a 62 and Khalid got a 65. I'm concerned about Jay, who usually does well. He got only a 49.

____

____

____

____

____

____

Stem-and-Leaf Plot Use the blank stem-and-leaf plot below to show the information on page 76. Then write one or two sentences describing what the plot shows about the results of this test.

Stem	Leaves
4	________
6	________
7	________
8	________
9	________
10	________

About this test

____

____

____

____

____

____

Compare and Contrast Diagrams

In this section, we have looked at four organizers you can use to communicate mathematical information. However, there are many other ways to communicate information. Here is an organizer you could use to compare and contrast different things. Experiment with different ways to set up graphic organizers to find the ones that work best for you.

Write the things you are comparing on the lines at the top. In the first box, say how they are alike. Write the characteristics you are contrasting under "With Regard To." Then show how the items are different according to these characteristics.

Item 1 ____________________________ Item 2 ____________________________

How Alike?

How Different?

With Regard to

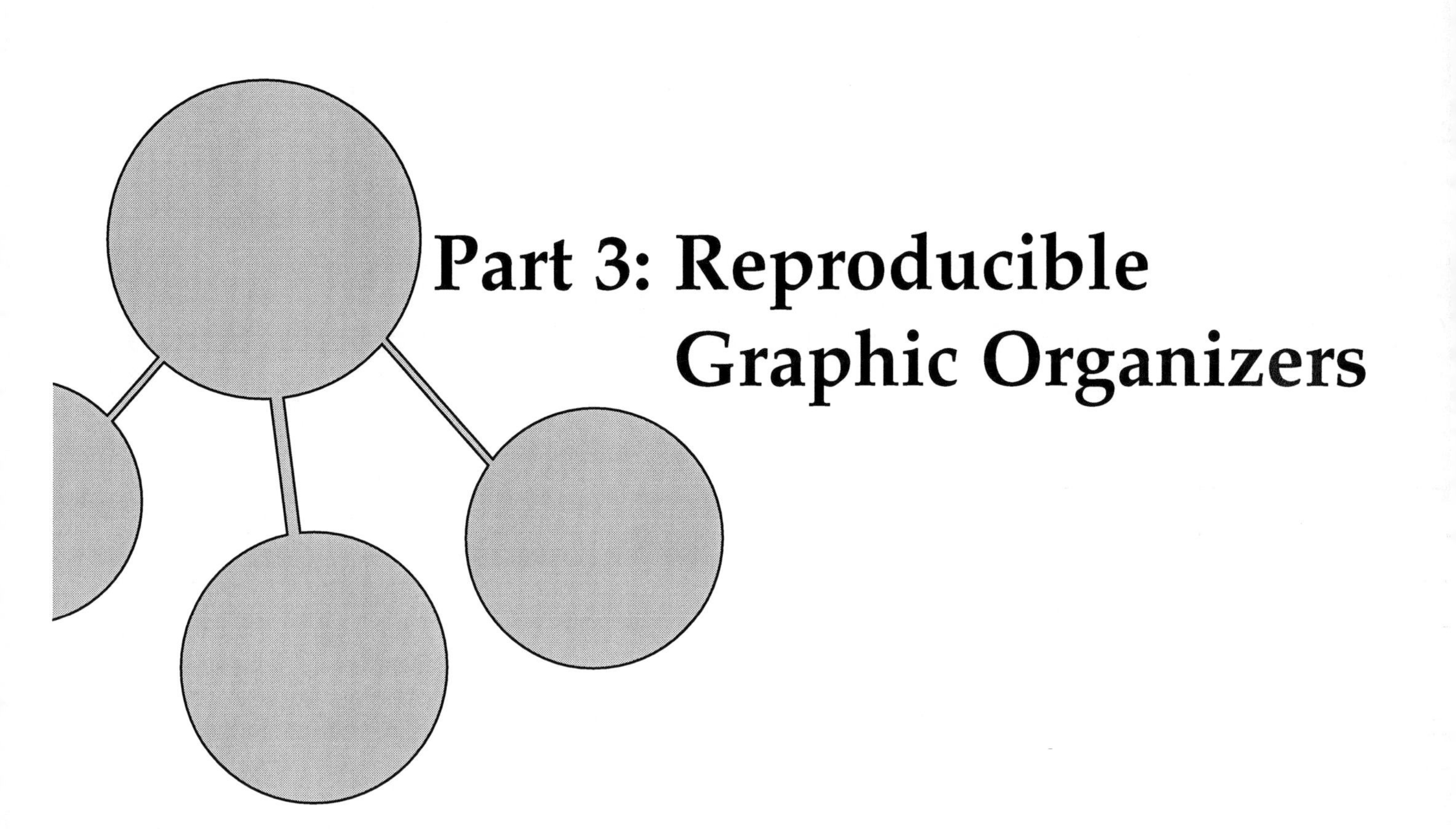

Part 3: Reproducible Graphic Organizers

Table

Flowchart

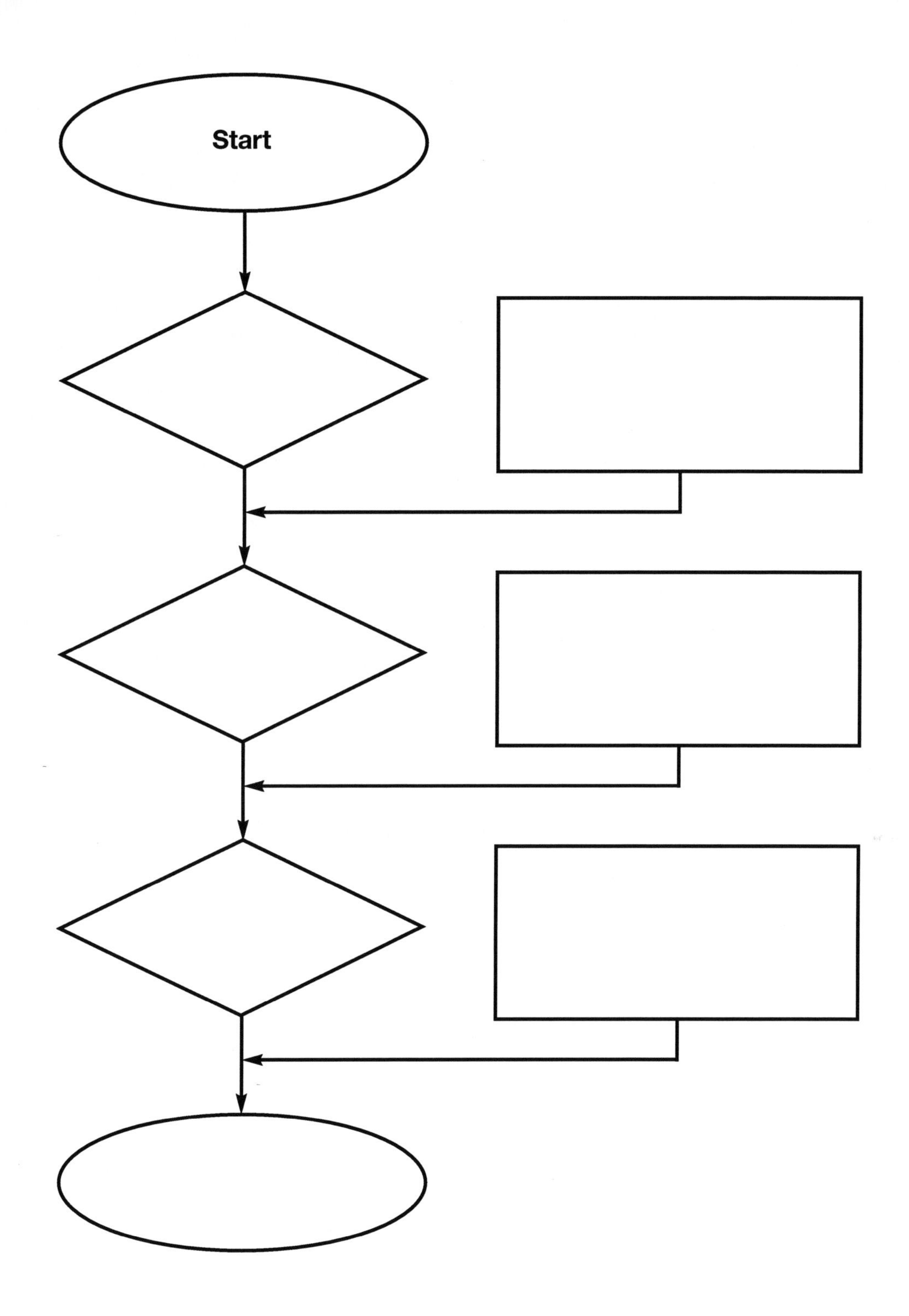

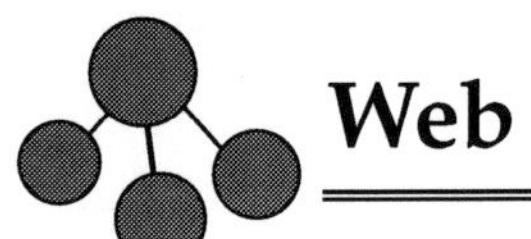

Web

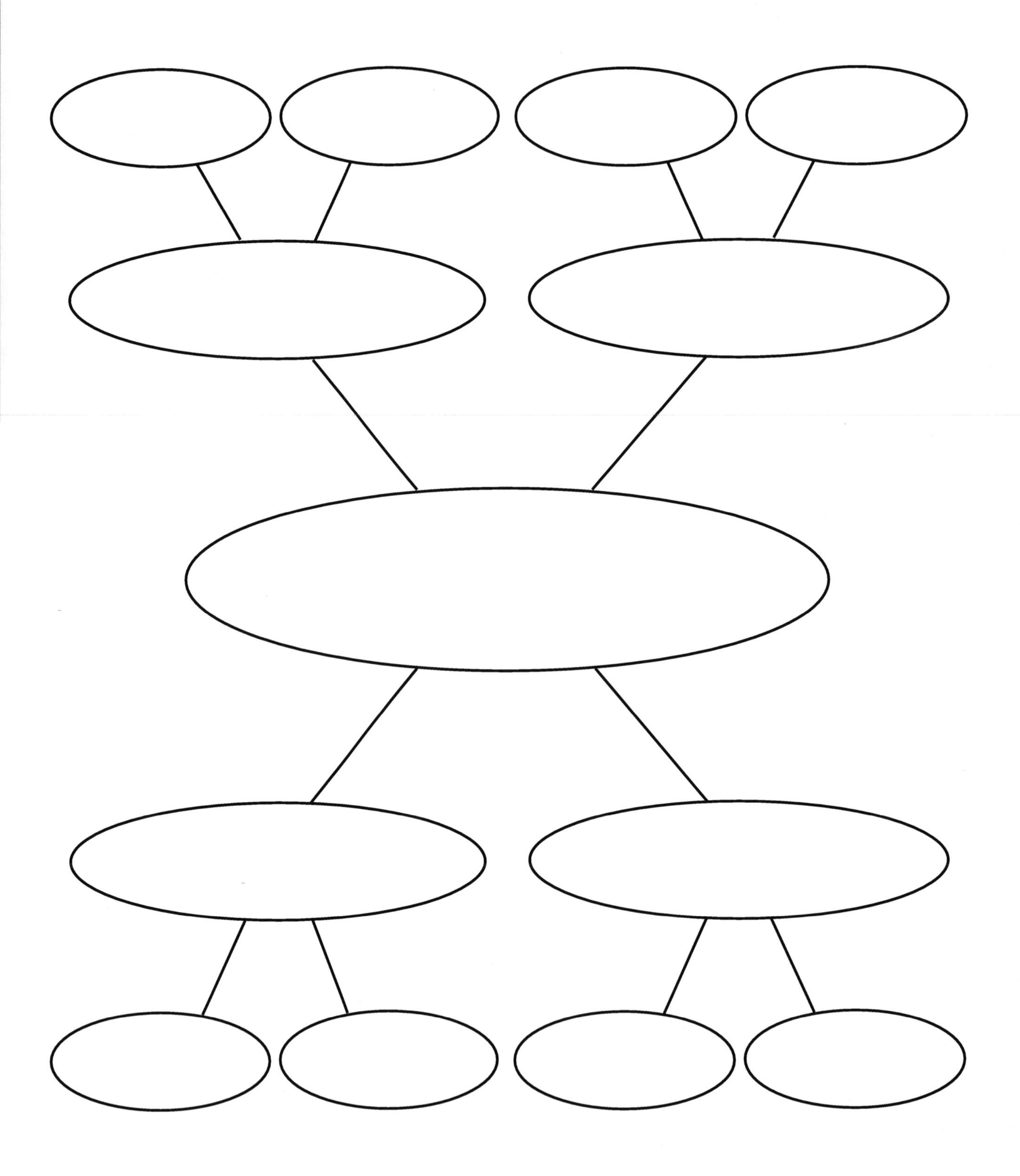

Number Line

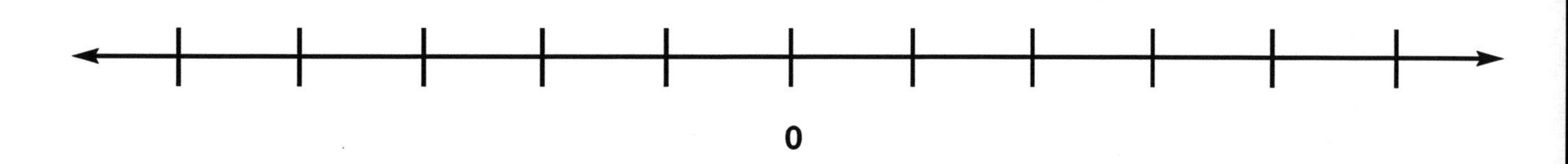

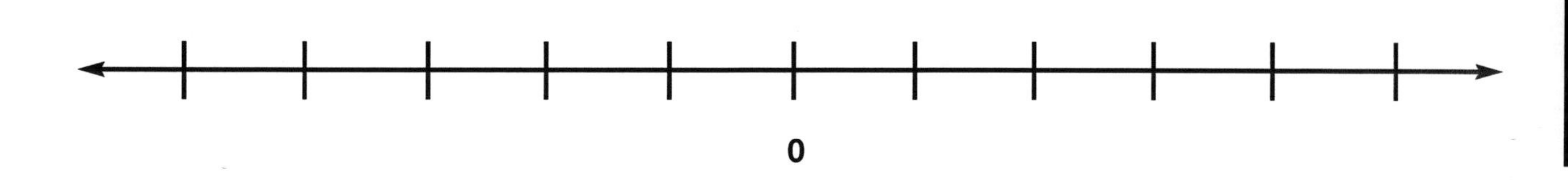

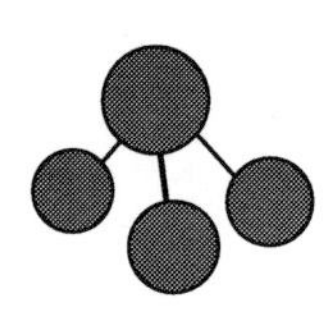

Geometric Drawing

Factor Tree

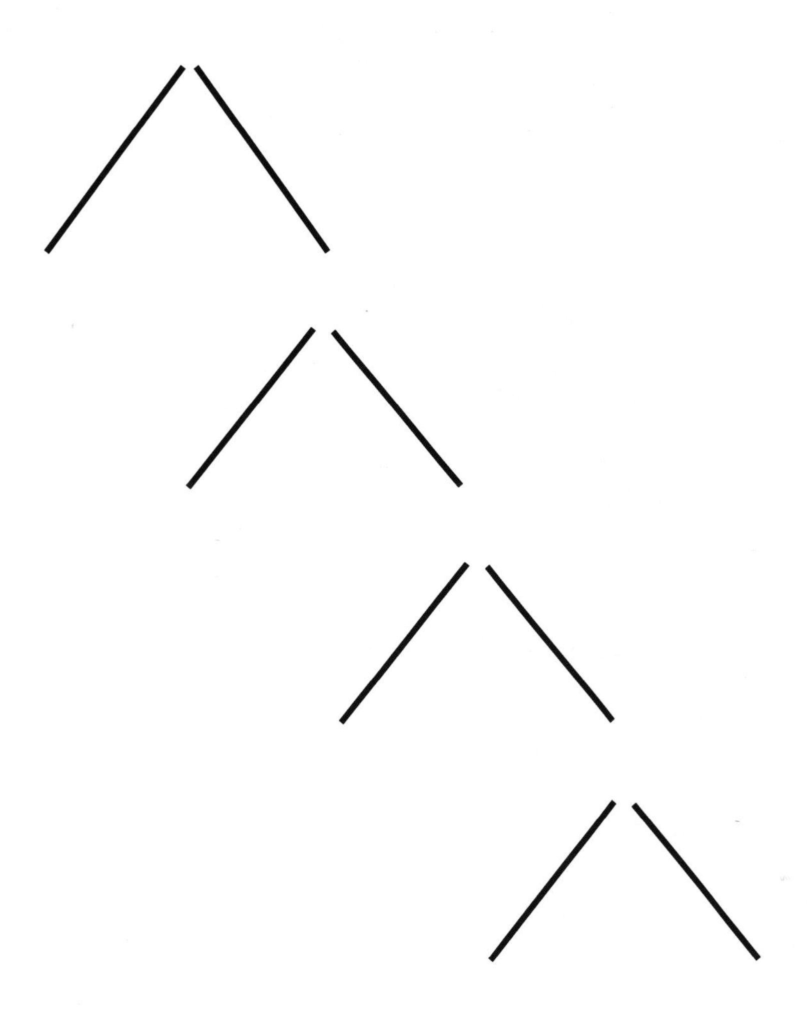

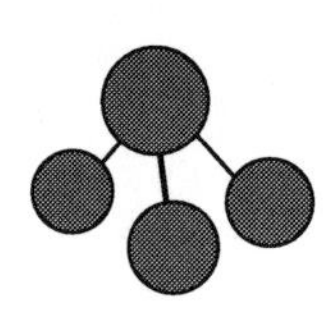

Venn Diagram

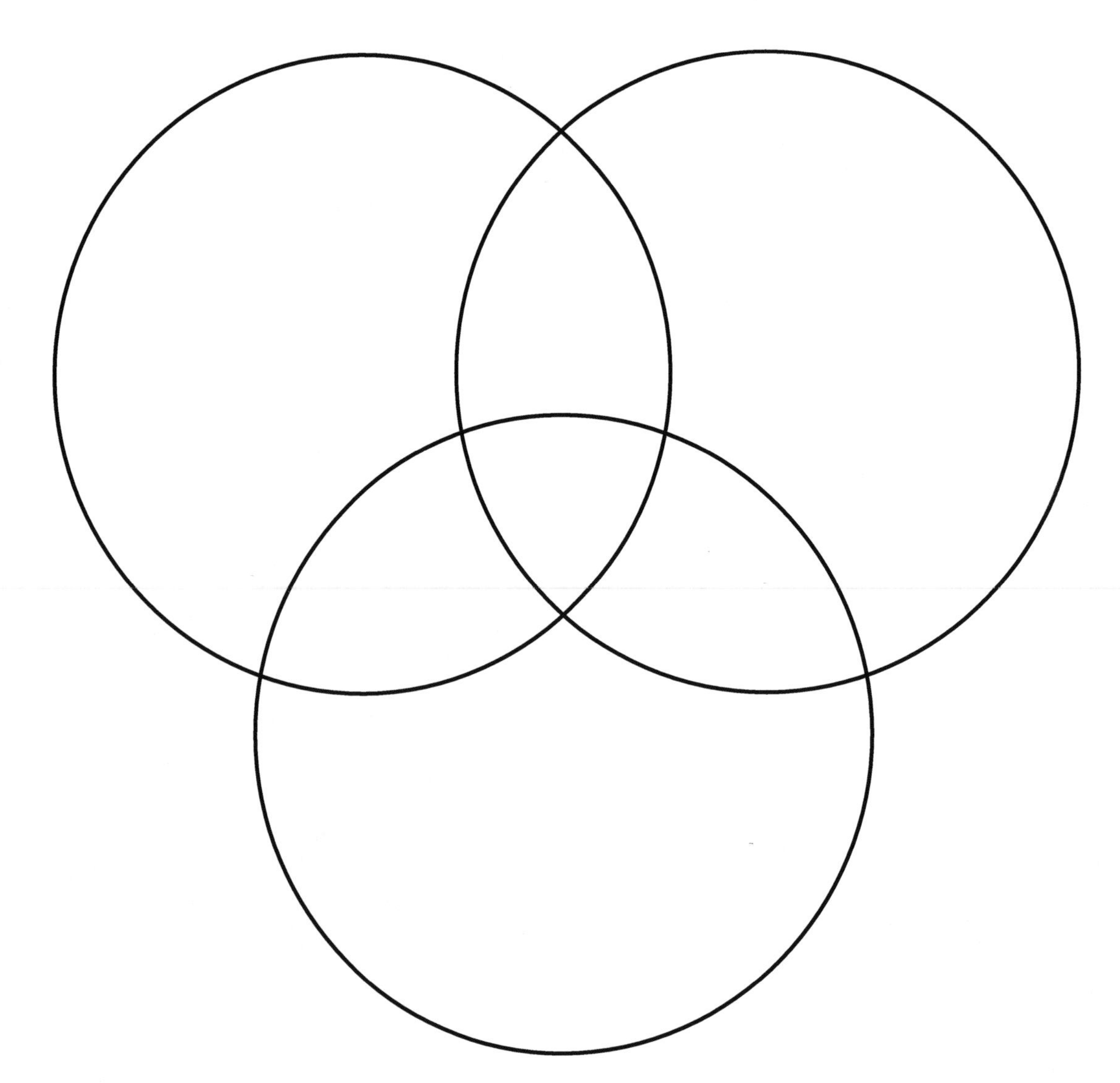

Probability Trees

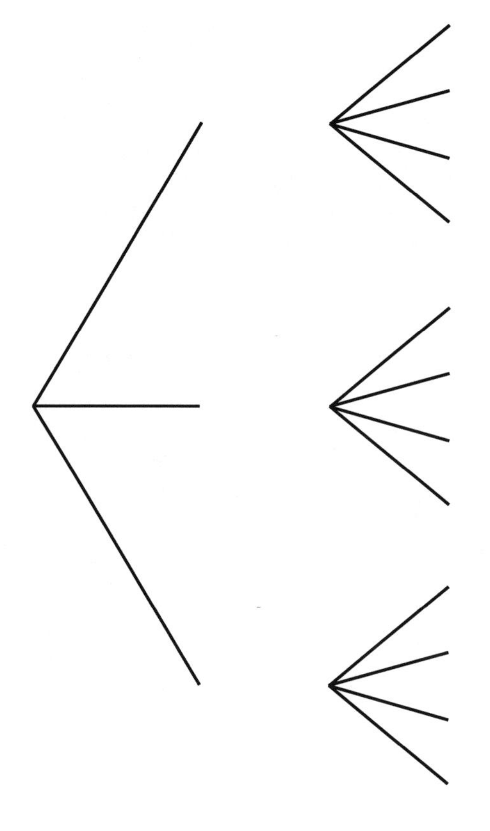

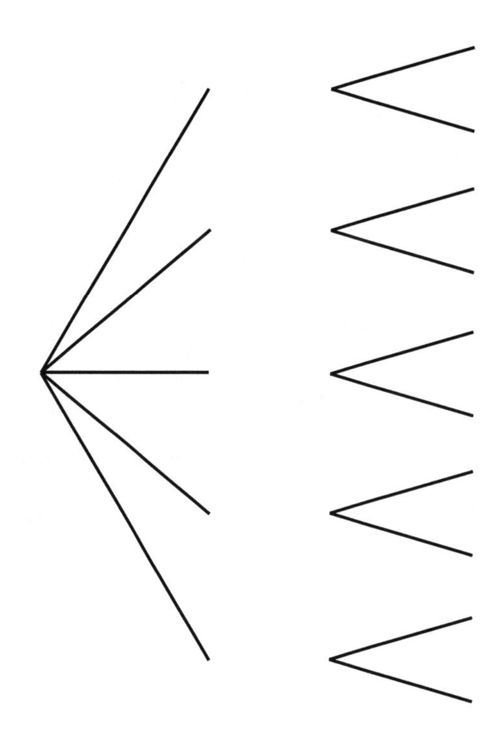

Attribute Table

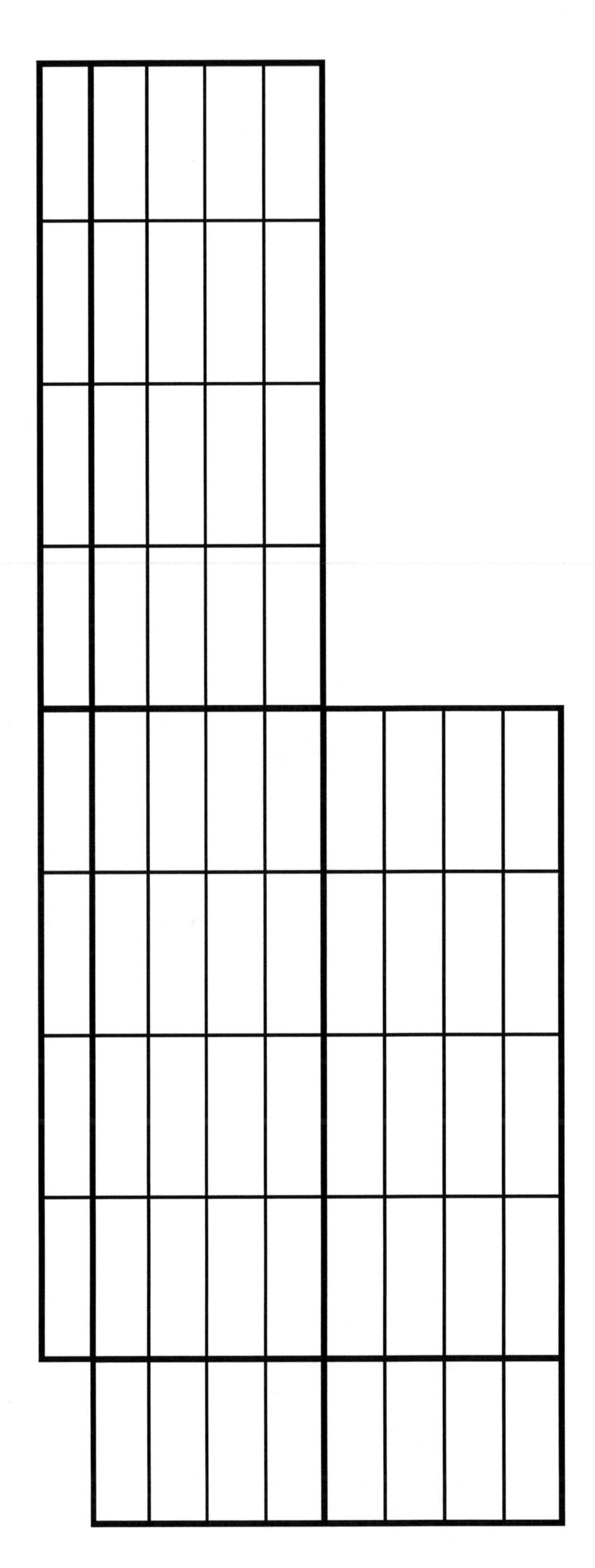

Cause and Effect Map

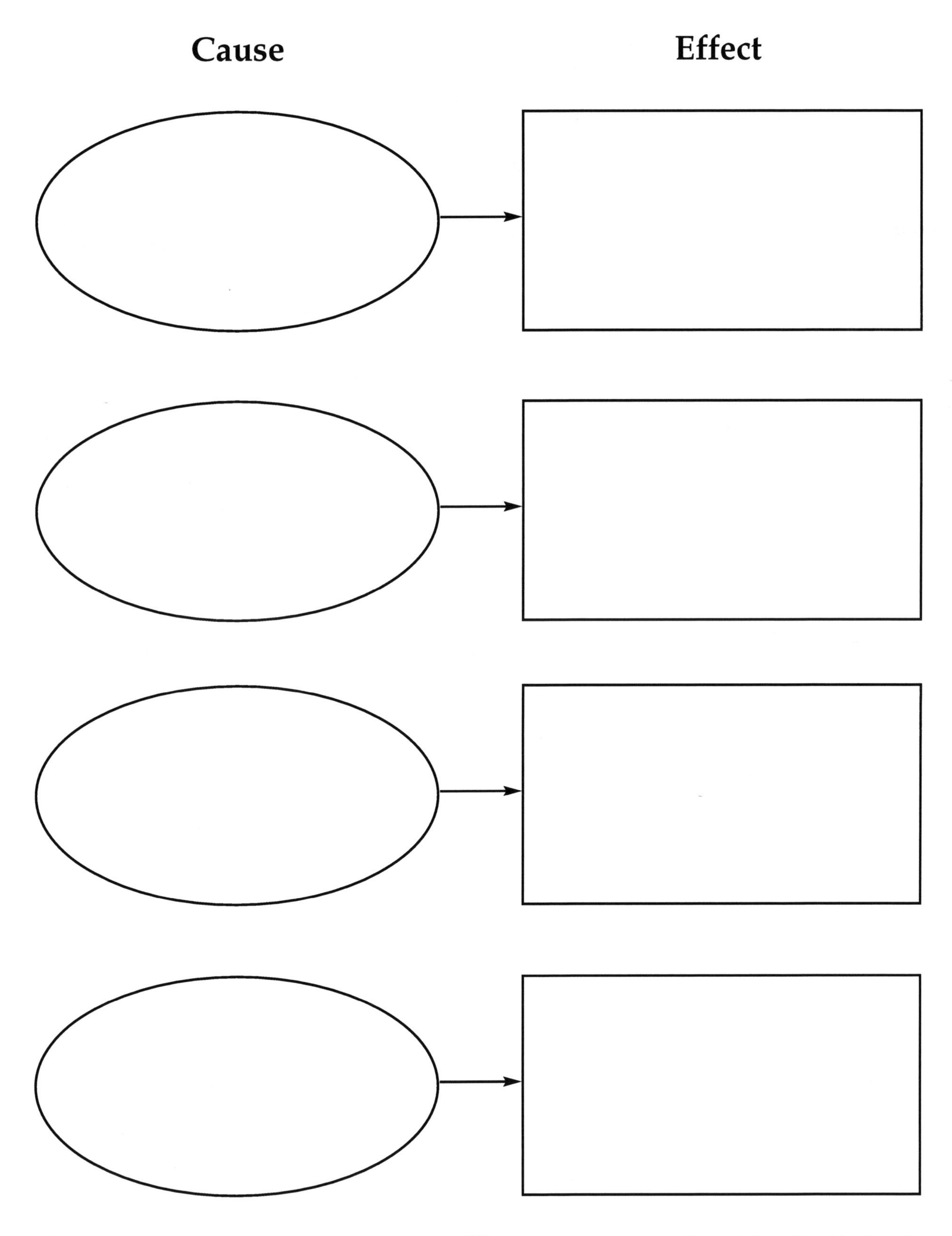

Line Graph

Graph title ________________________

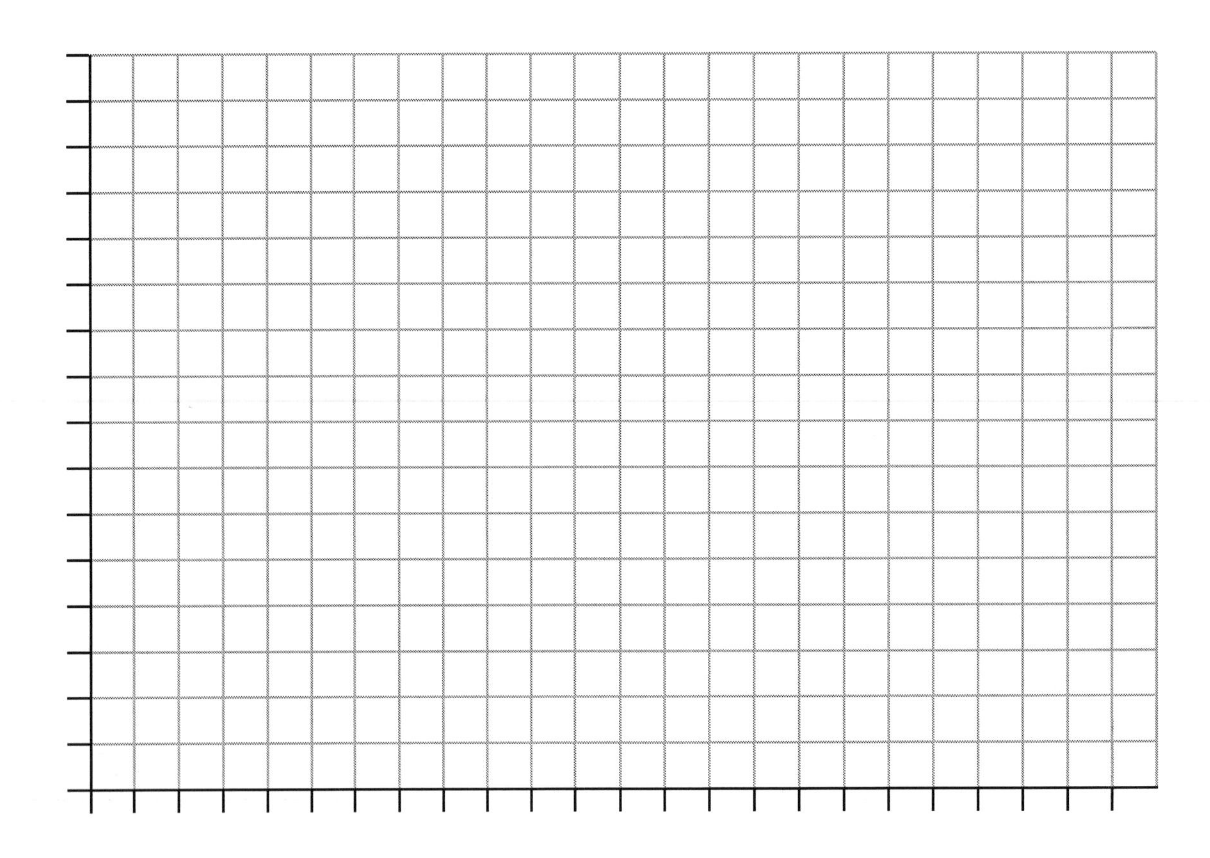

Axis title ________________

Axis title ____________________

Bar Chart

Graph title ______________________

Axis title ______________________

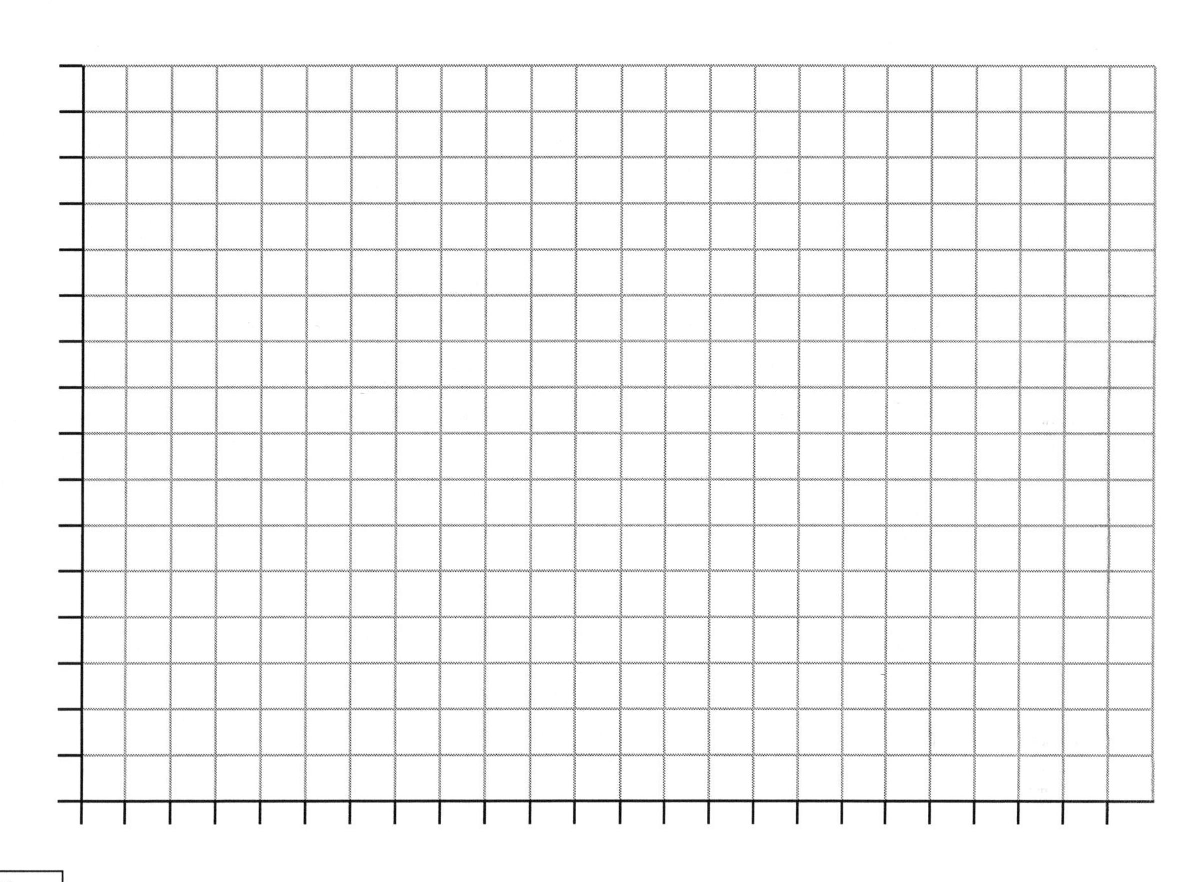

KEY
☐
☐
☐
☐

Axis title ____________________

Pie Chart

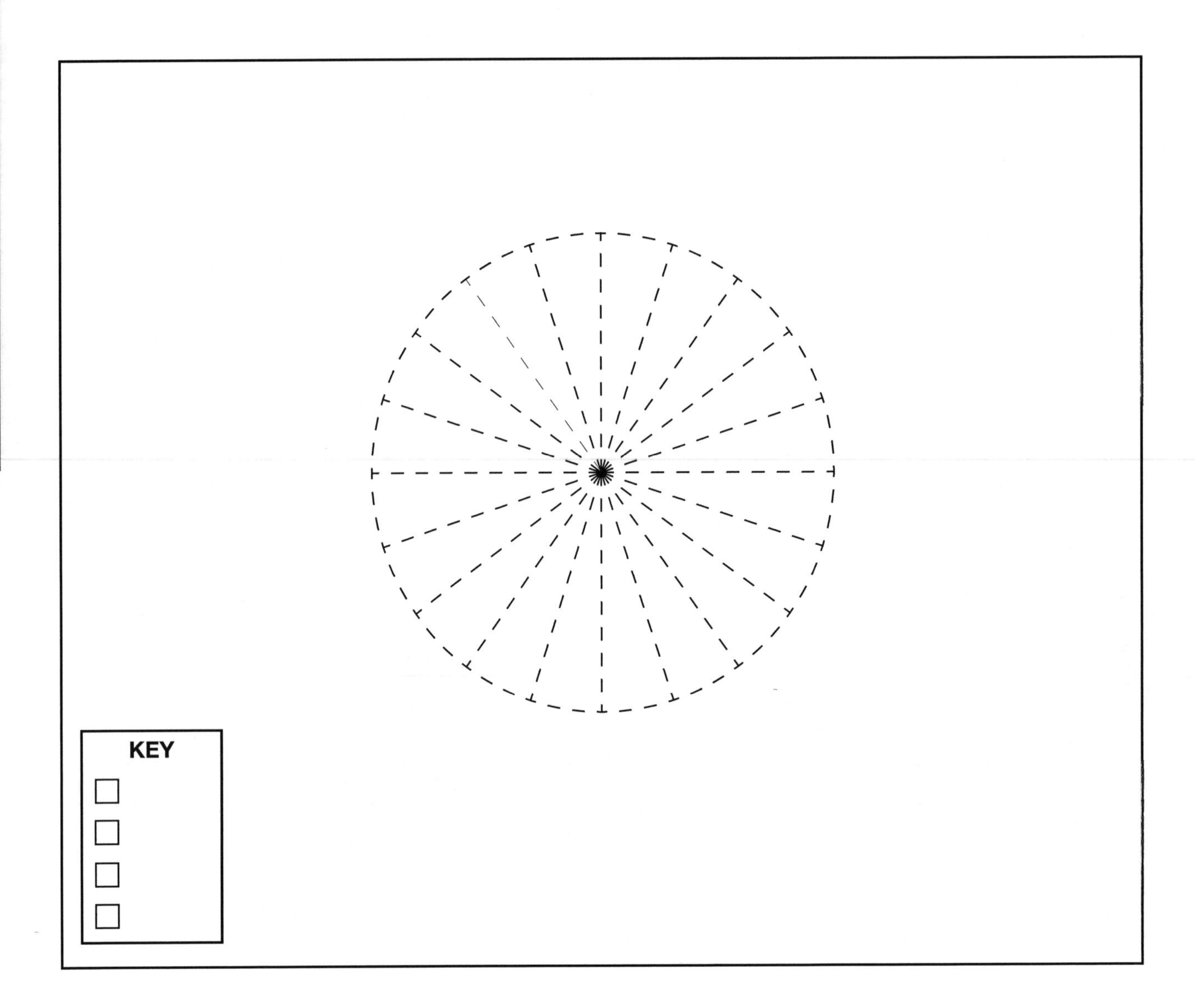

Stem-and-Leaf Plot

Stem	Leaf
___	______________
___	______________
___	______________
___	______________
___	______________
___	______________
___	______________

Leaf	Stem	Leaf
______________	___	______________
______________	___	______________
______________	___	______________
______________	___	______________
______________	___	______________
______________	___	______________
______________	___	______________

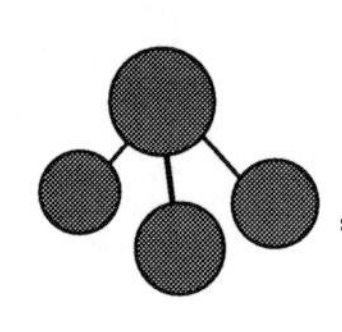

Compare and Contrast Diagram

Item 1 ______________________ Item 2 ______________________

How Alike?

How Different?

With Regard to

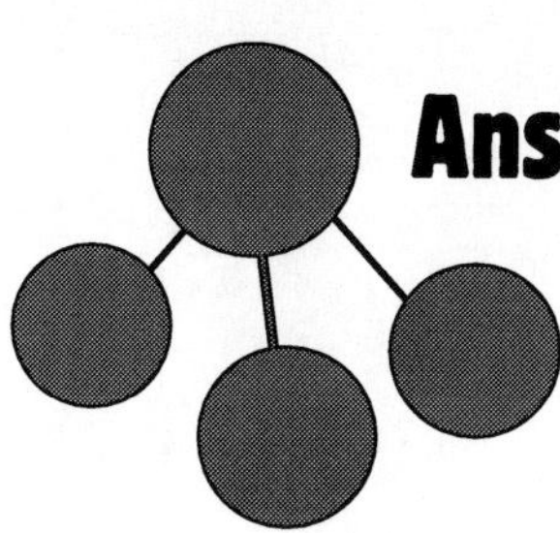

Answer Key: Lesson 2

Table, page 13

Polygon	Area Formula
Rectangle	$A = bh$
Square	$A = s^2$
Parallelogram	$A = bh$
Triangle	$A = \frac{1}{2}bh$
Trapezoid	$A = \frac{1}{2}(b_1 + b_2)h$

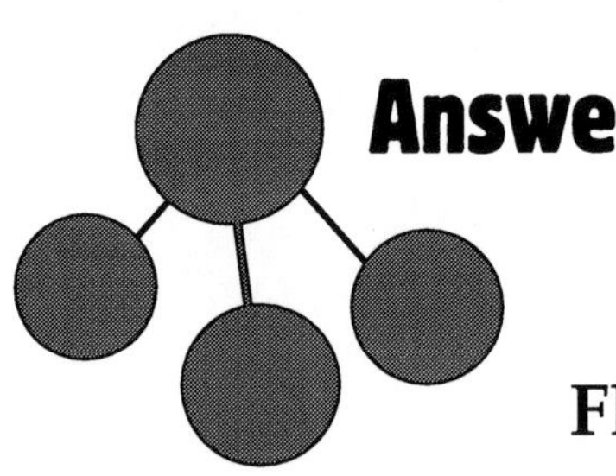

Answer Key: Lesson 2

Flowchart, page 18

Answers may vary. Sample answer:

Process for Evaluating Arithmetic Expressions

Start
Are there any parentheses in the expression?
If yes, do any calculations inside parentheses.
If no, are there any multiplication or division operations?
If yes, do multiplication and division, working from left to right.
If no, are there any addition and subtraction operations?
If yes, do addition and subtraction, working from left to right. Then stop.
If no, Stop.

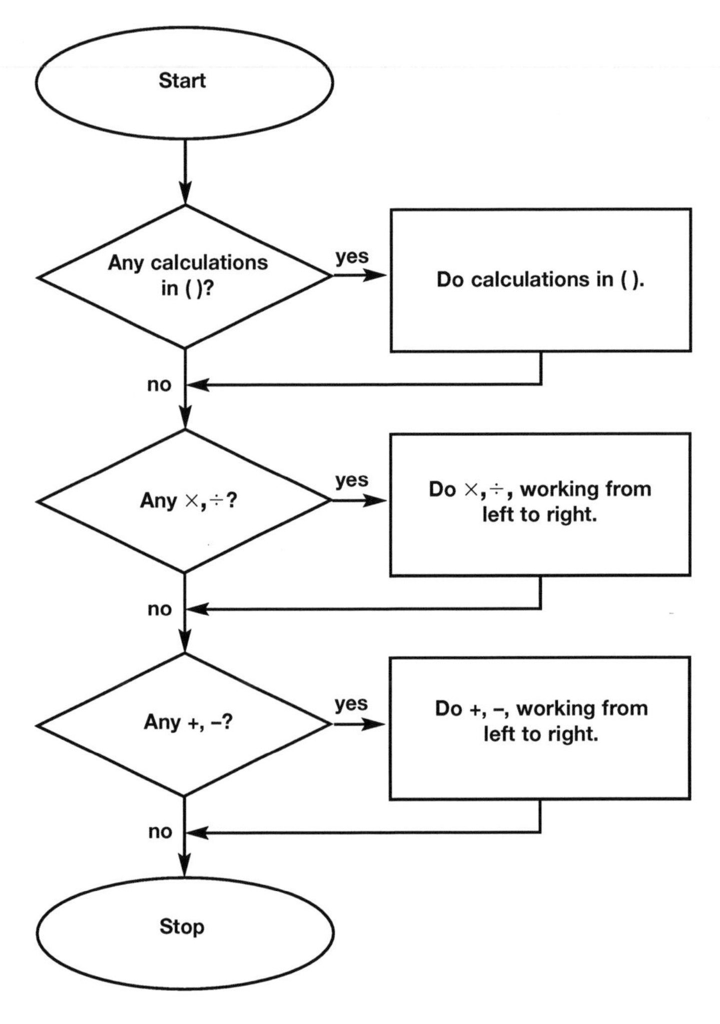

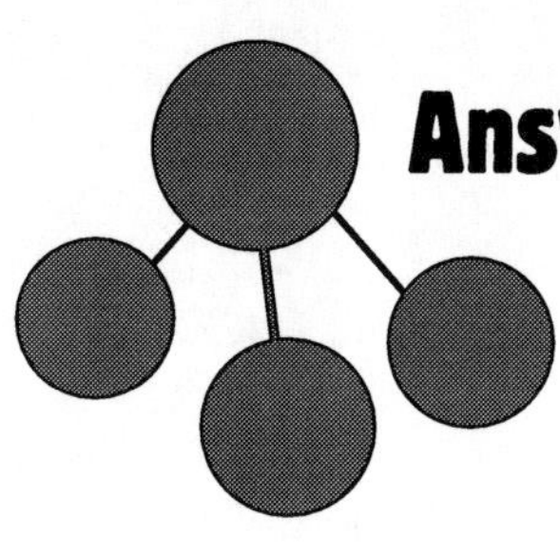

Answer Key: Lesson 3

Number Lines, page 25

1. Problem statements may vary. Sample statement: What were the maximum temperature, minimum temperature, and temperature range on January 8?

   Answer: Maximum: 4° F; minimum: –12° F; range: 16° F

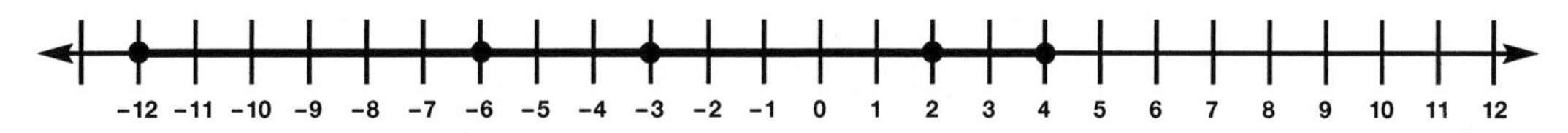

11 P.M. 7 P.M. 7 A.M. 3 P.M. 11 A.M.

2. Problem statements may vary. Sample statement: How many of 265 people would have to take later flights before 5 people could be seated on a 256-seat plane?

   Answer: 14 people

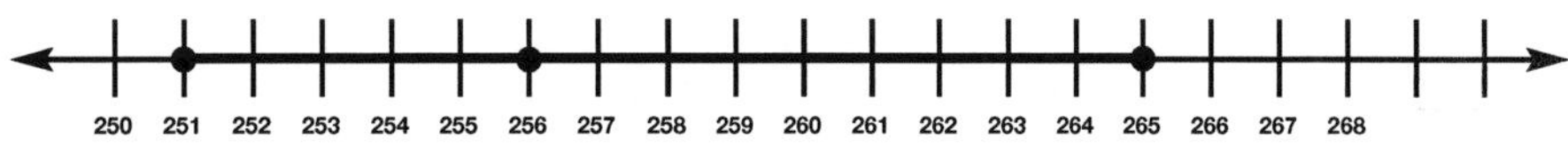

point at which Lees could get on flight

number of possible seats

number of booked seats

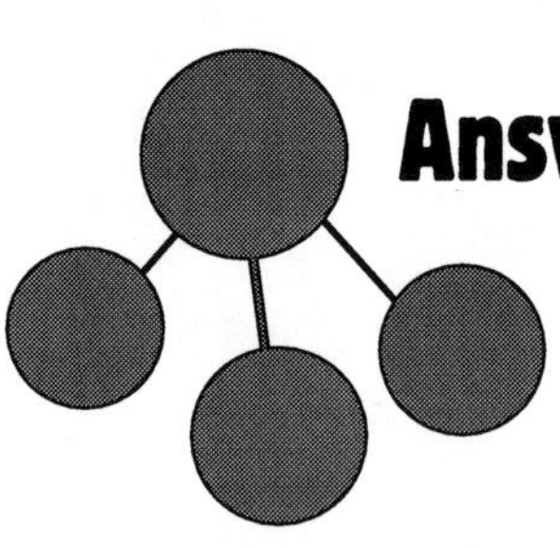

Answer Key: Lesson 3

Geometric Drawing, page 29

Student descriptions of the process will vary. Sample answer: I drew height lines (dotted), forming a rectangle. Then I drew line *a* parallel to the left side of the trapezoid, in effect copying the triangle on the left side of the trapezoid over next to the triangle on the right side. The area of the trapezoid is the sum of the area of the rectangle and the area of the larger triangle formed by *a*, the right side of the trapezoid, and the base. The area of the rectangle is 5 cm × 7 cm = 35 cm². The area of the larger triangle is $\frac{1}{2}$ × 5 cm × (15 cm – 7 cm) = 20 cm². The area of the trapezoid is 35 cm² + 20 cm² = 55 cm².

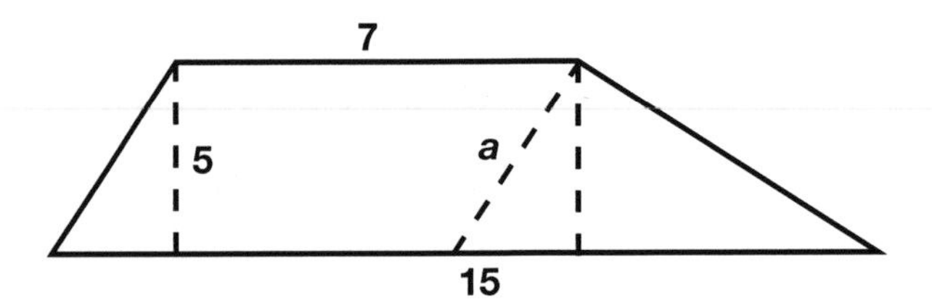

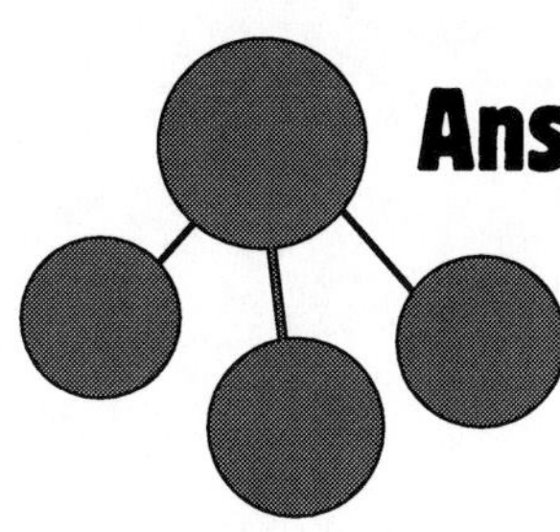

Answer Key: Lesson 3

Factor Trees, page 34

Student factor trees may vary. Sample answers:

1. $693 = 3^2 \times 7 \times 11$

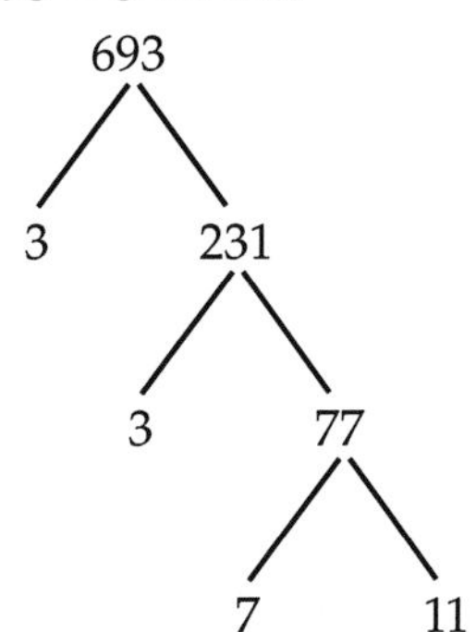

2. $1100 = 2^2 \times 5^2 \times 11$

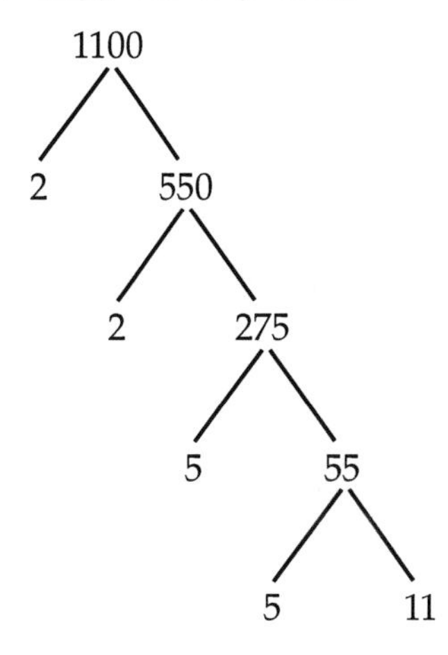

3. Prime factors of 48: $2 \times 2 \times 2 \times 2 \times 3$, or $2^4 \times 3$

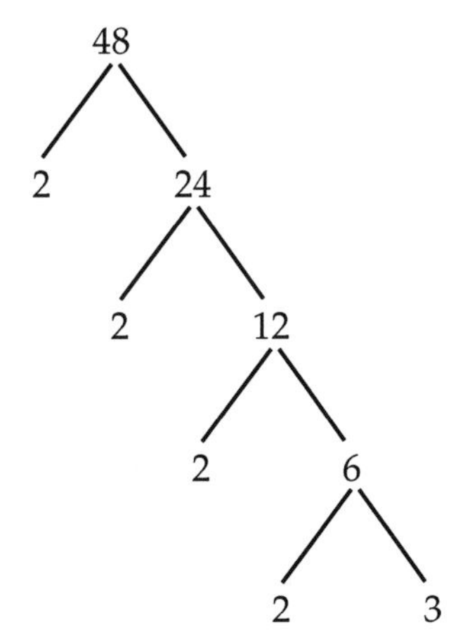

Prime factors of 72: $2 \times 2 \times 2 \times 3 \times 3$, or $2^3 \times 3^2$; GCF of 72 and 48 is $2^3 \times 3$ or 24

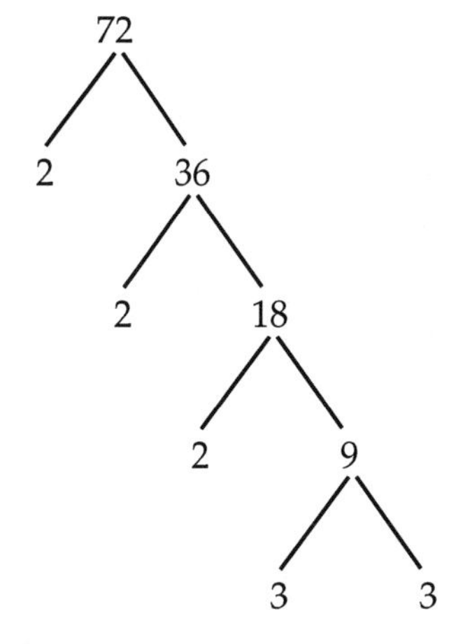

4. Prime factors of 540: $2 \times 2 \times 3 \times 3 \times 3 \times 5$, or $2^2 \times 3^3 \times 5$

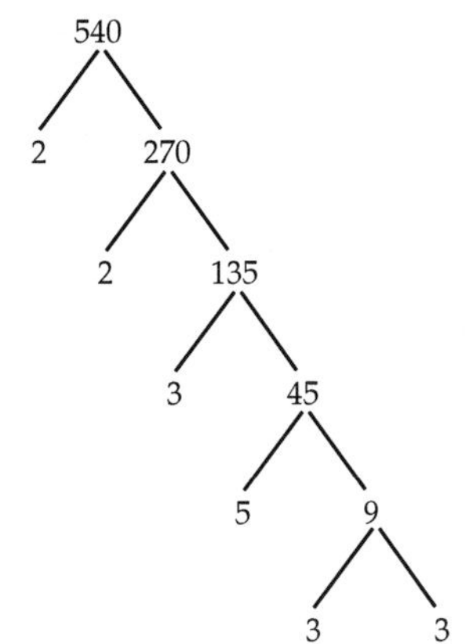

Prime factors of 126: $2 \times 3 \times 3 \times 7$, or $2 \times 3^2 \times 7$; GCF of 540 and 126 is $2 \times 3^2 = 18$

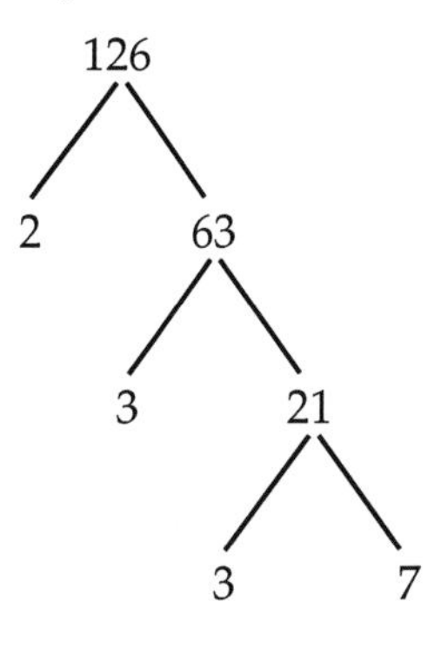

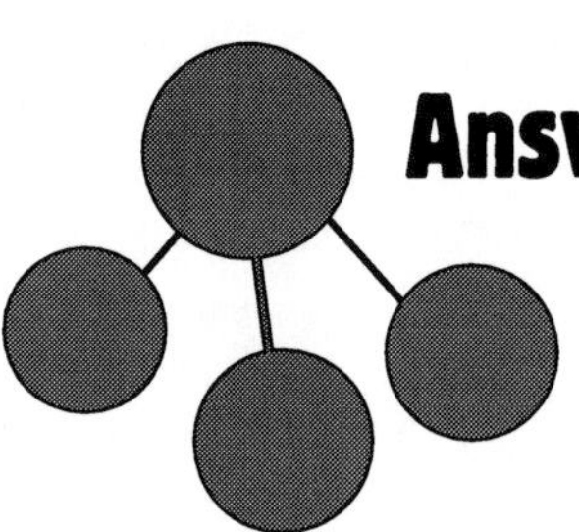

Answer Key: Lesson 3

Venn Diagrams, page 39

1. Three students read all three magazines. Fifteen read only one magazine. Student diagrams may vary. Sample diagram:

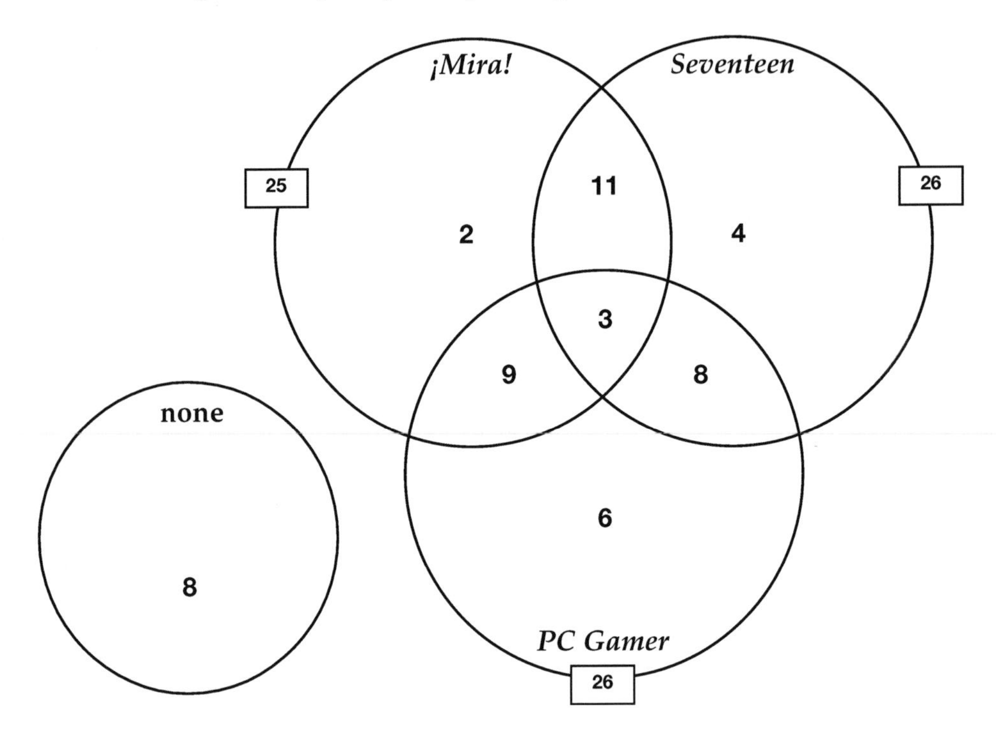

2. No, all vons are not shoshniks. Some are rizms. Student diagrams may vary. Sample diagram:

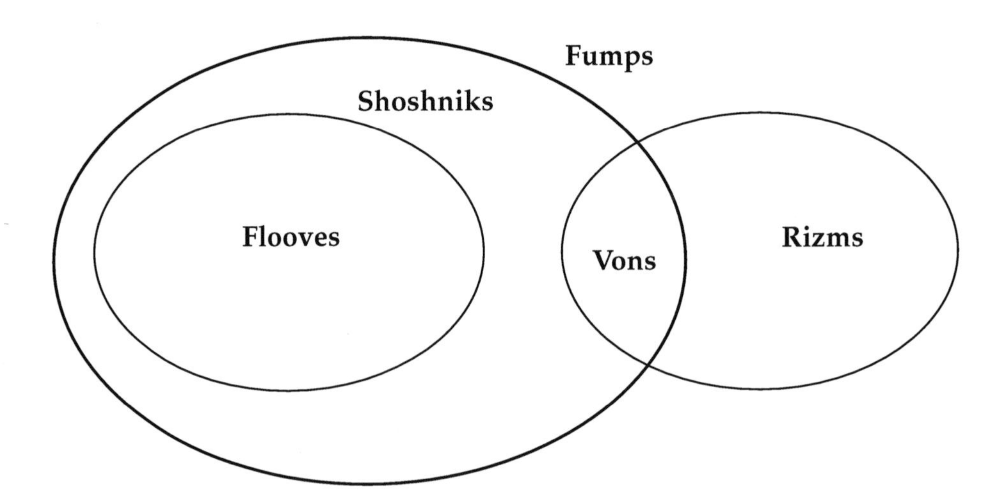

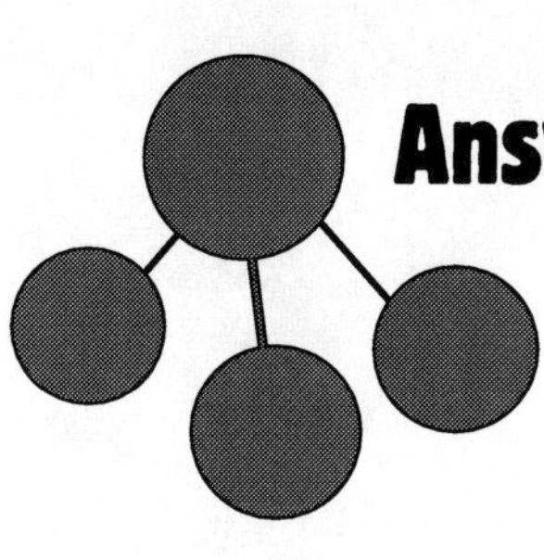

Answer Key: Lesson 3

Probability Trees, page 44

1. **a.** There are 16 possible outcomes. **b.** The probability that all four coins will land heads-up is $\frac{1}{16}$. **c.** The probability that all four coins will land tails-up is $\frac{1}{16}$.

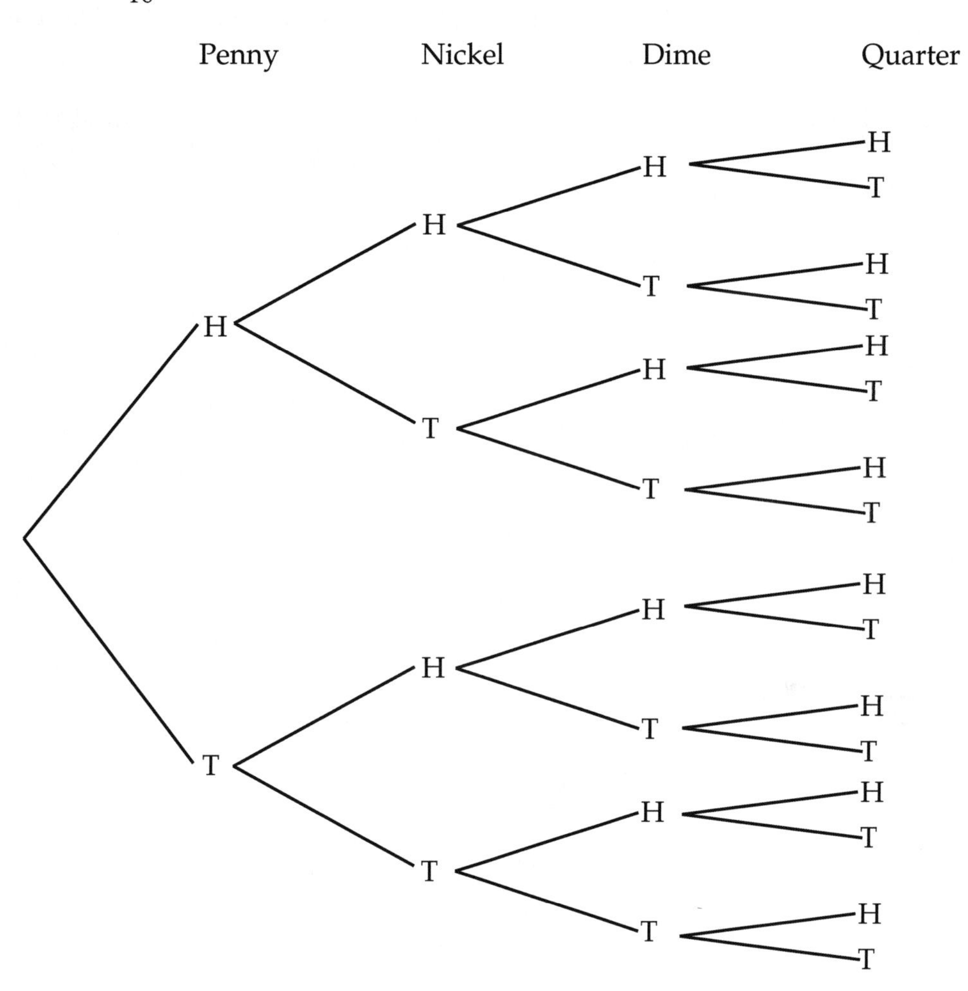

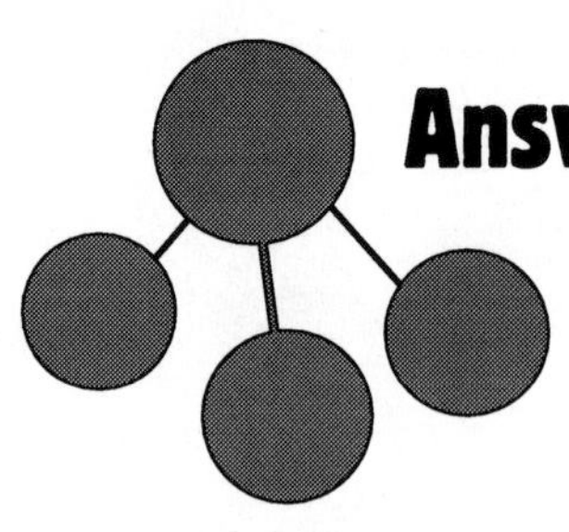

Answer Key: Lesson 3

Probability Trees, page 44, continued

2. **a.** There are 12 possible outcomes. (Note that although there are 7 players, only 6 are guests who will arrive at the door, fulfilling the requirements of the question.) **b.** The probability that the first guest to arrive will have reached Level 8 is $\frac{4}{12}$ or $\frac{1}{3}$. **c.** The probability that the last guest to arrive will have reached Level 7 is $\frac{2}{12}$ or $\frac{1}{6}$.

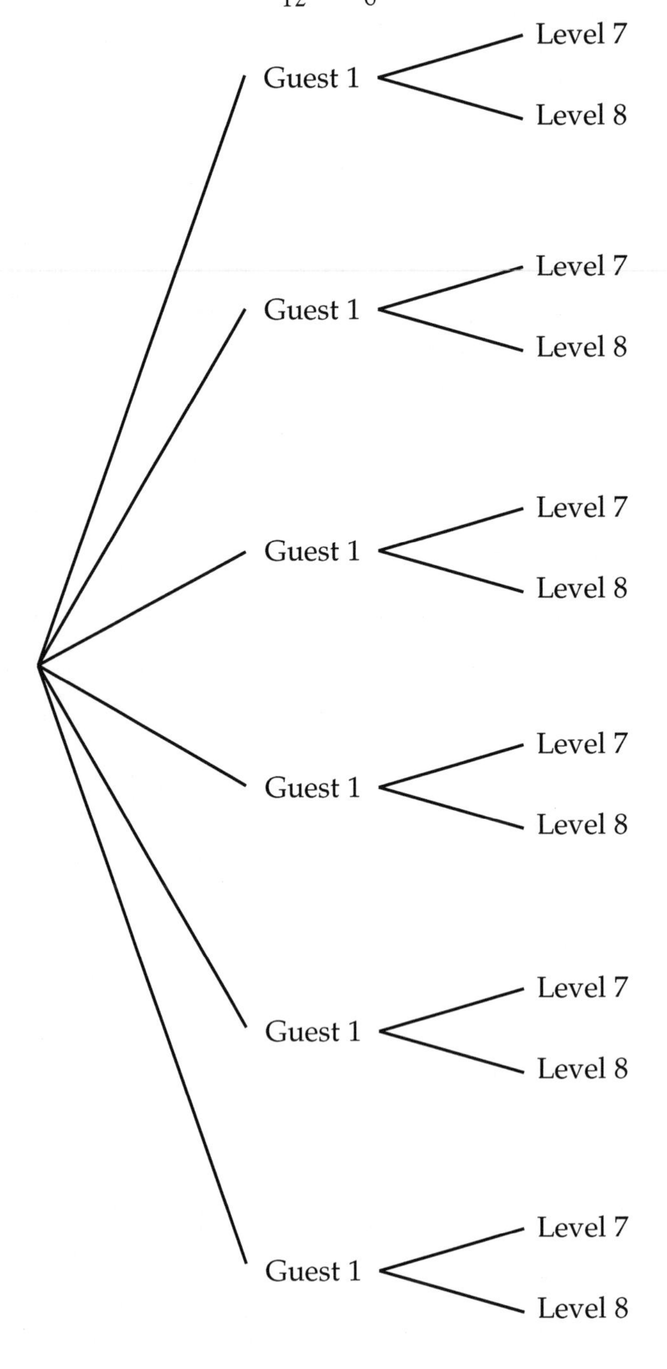

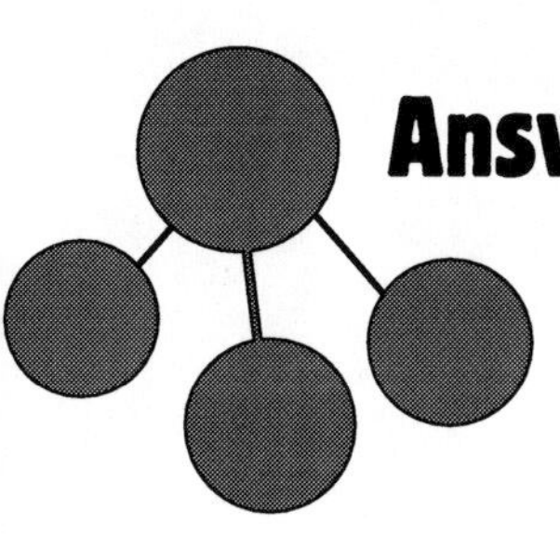

Answer Key: Lesson 3

Attribute Tables, page 50

1. Nicky plays soccer, Anna skis, and Kit plays hockey.

	soccer	hockey	skiing
Nicky	✔	X	X
Anna	X	X	✔
Kit	X	✔	X

2. Ardith went out with Tyler on Thursday. Mari went out with Alan on Sunday. Iris went out with Kevin on Friday. Jen went out with Jordan on Saturday.

	Tyler	Alan	Kevin	Jordan	Thursday	Friday	Saturday	Sunday
Ardith	✔	X	X	X	✔	X	X	X
Mari	X	✔	X	X	X	X	X	✔
Iris	X	X	✔	X	X	✔	X	X
Jen	X	X	X	✔	X	X	✔	X
Thursday	✔	X	X	X				
Friday	X	X	✔	X				
Saturday	X	X	X	✔				
Sunday	X	✔	X	X				

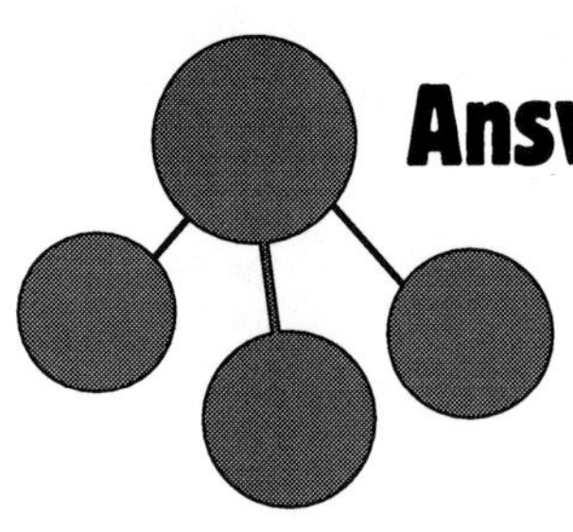

Answer Key: Lesson 4

Line Graph, page 57

1. At 0:05:00
2. At 0:18:00
3. 10.27 miles

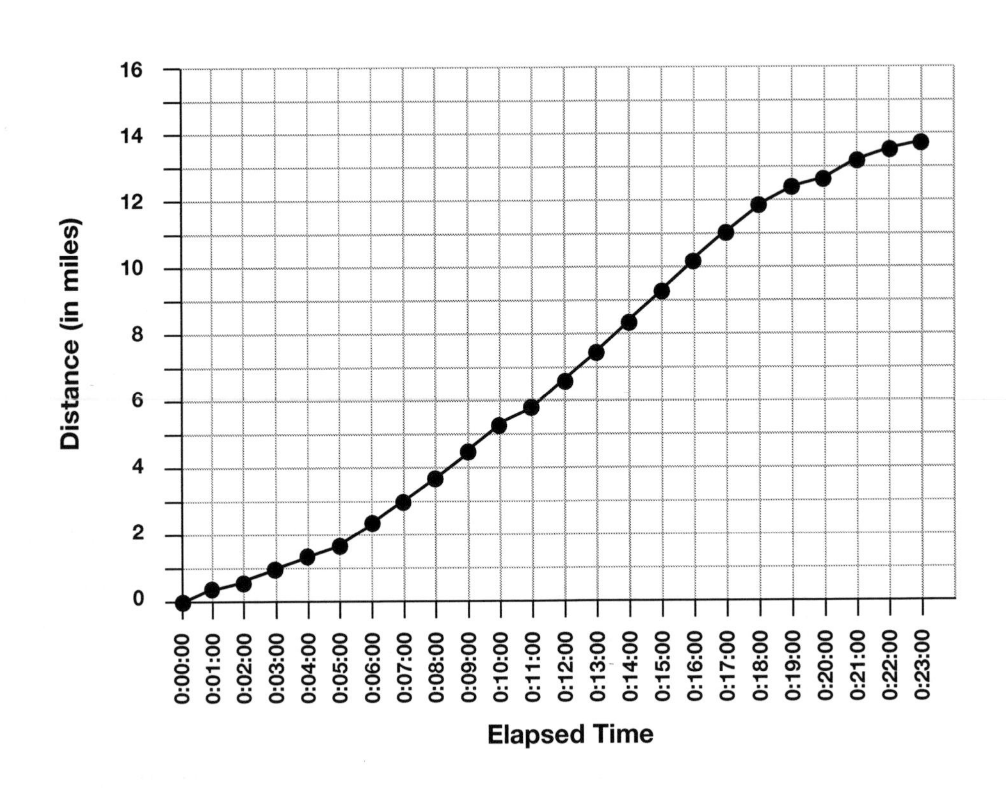

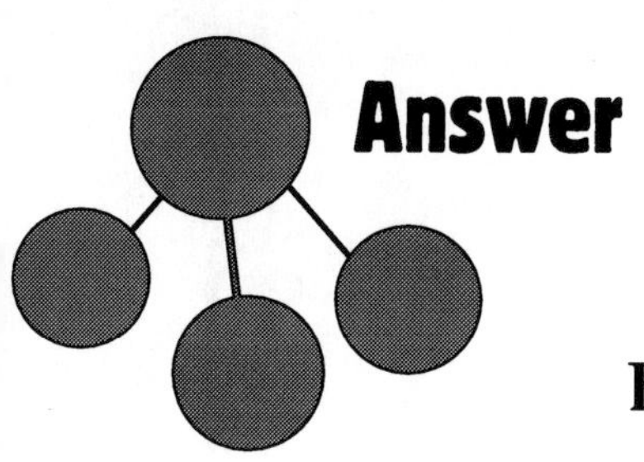

Answer Key: Lesson 4

Bar Chart, page 66

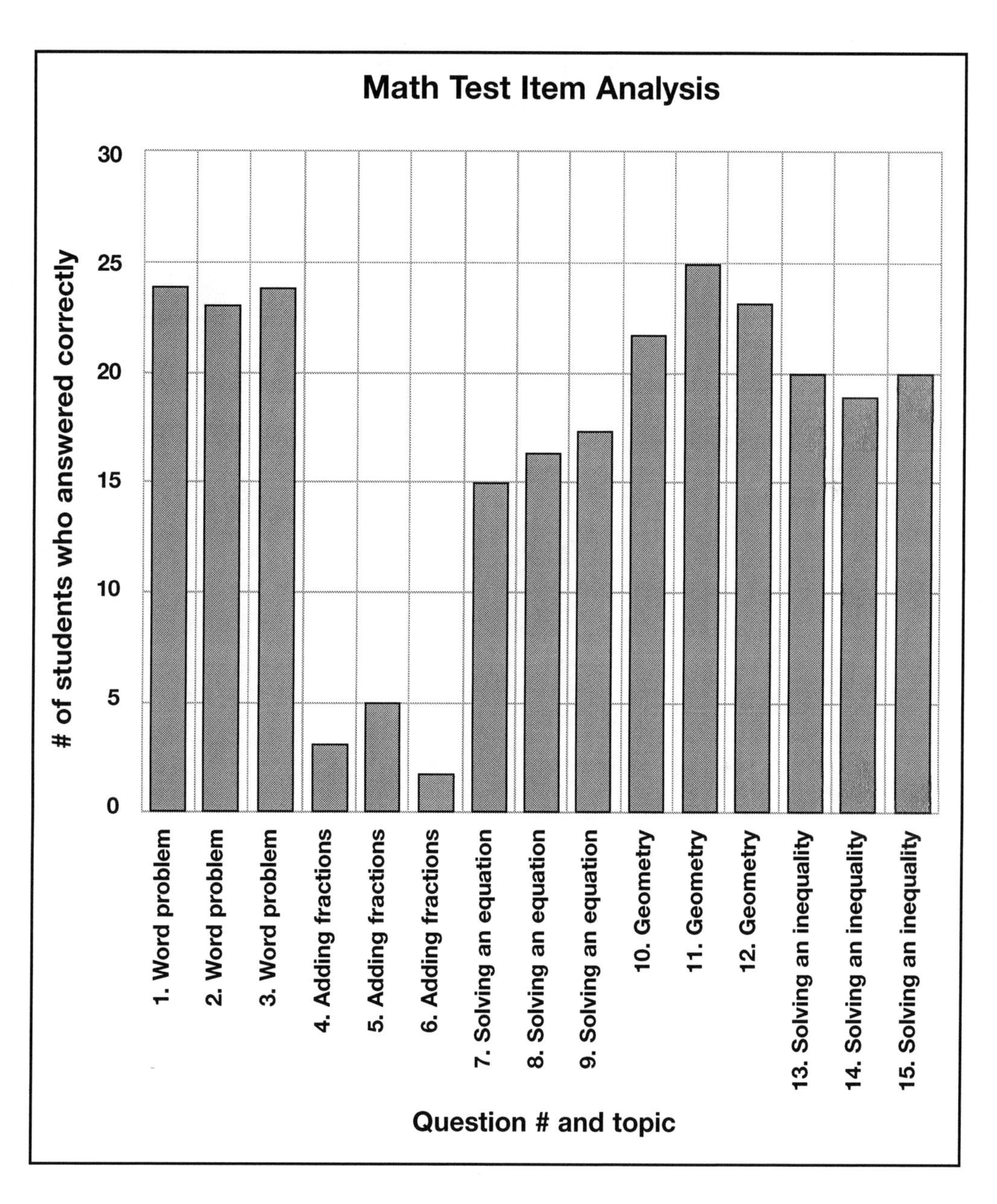

Answers will vary. Sample answer: Based on the chart, Ms. Washington should review fractions with her class.

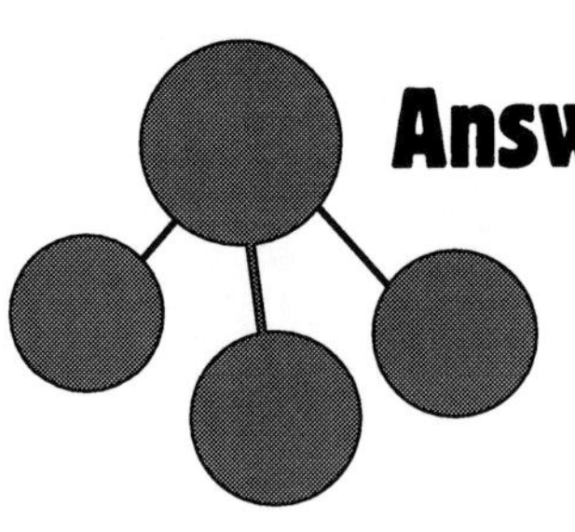

Answer Key: Lesson 4

Pie Charts, page 72

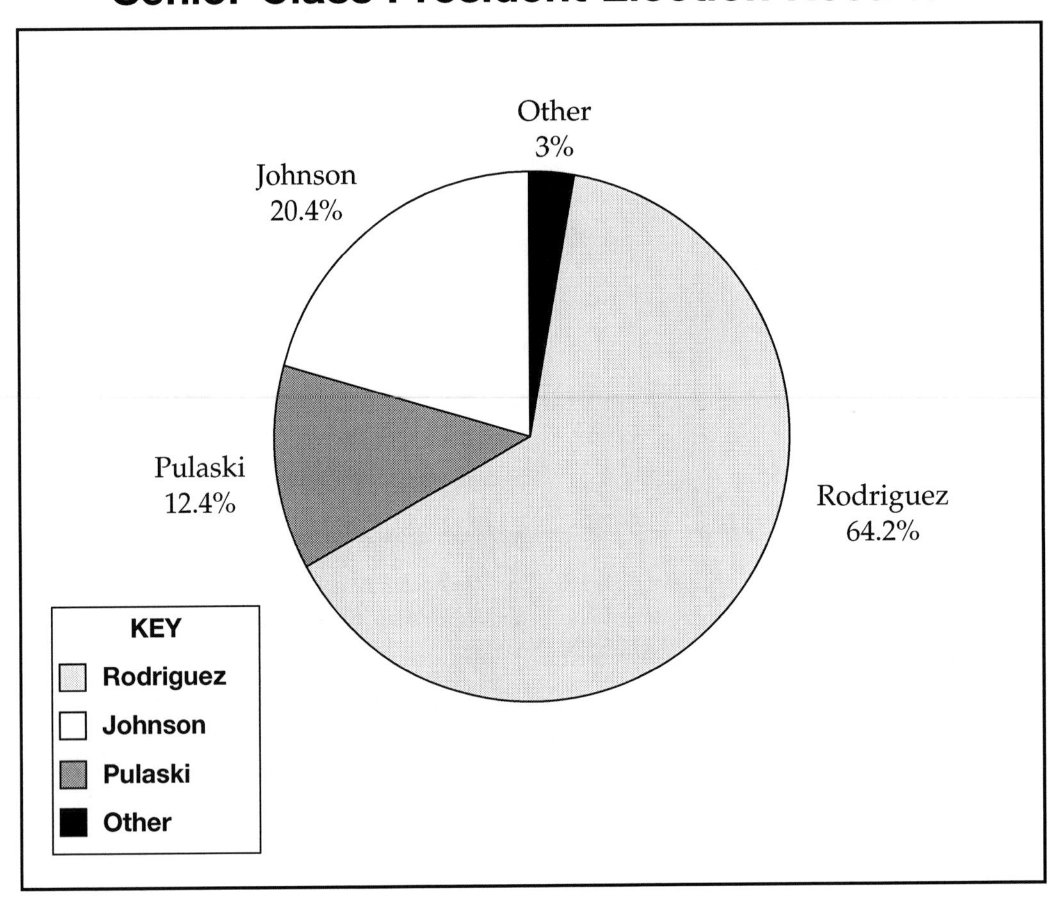

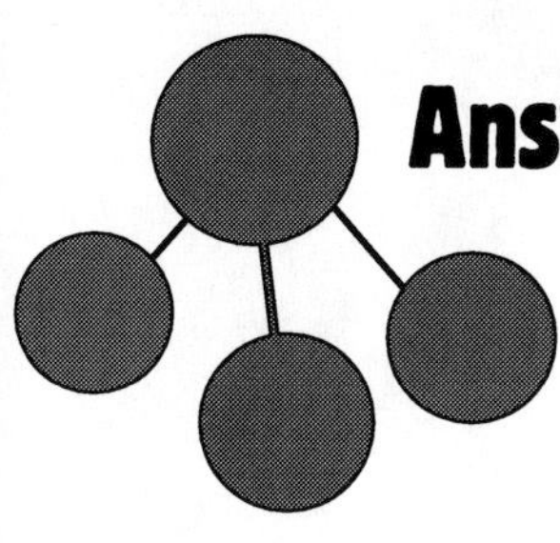

Answer Key: Lesson 4

Stem-and-Leaf Plot, page 77

Stem	Leaf
4	9
6	2 5
7	0 4 4 6 7
8	3 4 8 8
9	6 7
10	0

About this test: Answers will vary. Sample answer: The plot shows that most students in the class scored in the 70s and 80s. Only one student scored far below the rest of the class. Three students scored extremely well.

Share Your Bright Ideas

We want to hear from you!

Your name________________________________Date________________

School name____

School address____

City____________________State____Zip________Phone number (_____)__________

Grade level(s) taught________Subject area(s) taught________________________

Where did you purchase this publication?______________________________

In what month do you purchase a majority of your supplements?____________________

What moneys were used to purchase this product?

___School supplemental budget ___Federal/state funding ___Personal

Please "grade" this Walch publication in the following areas:

Quality of service you received when purchasing	A	B	C	D
Ease of use	A	B	C	D
Quality of content	A	B	C	D
Page layout	A	B	C	D
Organization of material	A	B	C	D
Suitability for grade level	A	B	C	D
Instructional value	A	B	C	D

COMMENTS:____

____

What specific supplemental materials would help you meet your current—or future—instructional needs?

____

Have you used other Walch publications? If so, which ones?________________________

May we use your comments in upcoming communications? ___Yes ___No

Please **FAX** this completed form to **888-991-5755**, or mail it to

Customer Service, Walch Publishing, P. O. Box 658, Portland, ME 04104-0658

We will send you a **FREE GIFT** in appreciation of your feedback. **THANK YOU!**